Military Service and the Integration of Jews into the Roman Empire

The Brill Reference Library of Judaism

VOLUME 72

The titles published in this series are listed at *brill.com/brlj*

Military Service and the Integration of Jews into the Roman Empire

By

Raúl González Salinero

BRILL

LEIDEN | BOSTON

Cover illustration: Sign from the Zodiac mosaic floor in the Hammat Tiberias synagogue (4th century CE). Photograph taken by Alan Avery-Peck in 2009.

Library of Congress Cataloging-in-Publication Data

Names: González Salinero, Raúl, author.
Title: Military service and the integration of Jews into the Roman empire / Raúl González Salinero.
Description: Leiden ; Boston : Brill, [2022] | Series: The Brill reference library of Judaism, 1571-5000 ; volume 72 | Includes bibliographical references and index.
Identifiers: LCCN 2021057961 (print) | LCCN 2021057962 (ebook) | ISBN 9789004506756 (hardback) | ISBN 9789004507258 (ebook)
Subjects: LCSH: Rome–History, Military–30 B.C.-476 A.D. | Jews–History, Military. | Jewish soldiers–Italy–Rome–History. | Rome–History, Military–Religious aspects. | Rome–Ethnic relations.
Classification: LCC U35 .G67 2022 (print) | LCC U35 (ebook) | DDC 355.009456/32–dc23/eng/20211220
LC record available at https://lccn.loc.gov/2021057961
LC ebook record available at https://lccn.loc.gov/2021057962

Typeface for the Latin, Greek, and Cyrillic scripts: “Brill”. See and download: brill.com/brill-typeface.

ISSN 1571-5000
ISBN 978-90-04-50675-6 (hardback)
ISBN 978-90-04-50725-8 (e-book)

This book is printed on acid-free paper and produced in a sustainable manner.

In memory of my beloved brother Rafael Juan

Contents

Preface

This book is the result of a project that stretches back in time. Shortly after publishing my research paper "El servicio militar de los judíos en el ejército romano" [Jewish Military Service in the Roman Army] in 2003, I realized that many of the subjects I covered in it would need some revision and that there was growing interest in the subject. That is why, in spite of having received several gracious and generous offers to have that paper published in English—first from Jonathan P. Roth and, more recently, from Christopher B. Zeichmann—, I decided to revise and expand it in depth and to publish the results as a monograph. The time, however, was not ripe ...

Years later, I received a grant from the Spanish Ministry of Science, Innovation and Universities as part of its study stay programme for senior professors and researchers in foreign institutions of higher learning and research, which is included in the Salvador de Madariaga programme. As a result, I had a research stay at the Faculty of Divinity of the University of Cambridge as a visiting fellow in Wolfson College, and I am grateful to its President, Professor Jane Clarke, for her warm welcome and for having put the resources of the College, whose staff was ever pleasant and efficient, at my disposal. The time had finally come for me to make inroads into my long-awaited project on the presence of Jewish soldiers in the Roman army, and I had the chance to present part of its results to the *Hebrew, Jewish and Early Christian Studies Senior Seminar*, directed by Professors William Horbury, Nicholas de Lange and James Carleton Paget. The exchange of ideas and suggestions during this Seminar allowed me to make significant improvements to my research. In addition to its convenors, I thank Irina A. Levinskaya (Russian Academy of Sciences) and Geoffrey Greatrex (University of Ottawa) for their thoughtful comments.

I never would have imagined that, for a number of reasons, this would be one of the hardest assignments in my academic career. My elder brother passed away during the first days of that stay, in the Michaelmas Term 2018, his will to live and the encouragements of his loved ones having been powerless in the face of death. For this reason, this book is dedicated to his memory, as my recollection of him, which was on my mind through every passing hour in the beautiful city of Cambridge, is indissolubly linked to it.

Among us the living, immersed in the daily struggle of our lives, ever full of problems and concerns, but also hope and joy, there are many people without whom this book would not have been possible. First of all, I would like to thank Professor Nicholas de Lange, my mentor and *magister* in Cambridge, from whom I have learnt so much, and whose humanity is without bounds. I still yearn for our endless conversations. Likewise, I cannot imagine a better host than Professor William Horbury, with whom I shared unforgettable moments at Corpus Christi College. His kindness and contagious joy are only matched by his awe-inspiring wisdom. Among the magnificent group from Cambridge, I must also mention Professors James Nicholas Carleton Paget, James K. Aitken, George van Kooten and Judith Lieu. It was an honour for me to have met them and for them to have shown an interest in my research. Until that moment, some of them had only been present in my academic dreams, but those dreams had finally come true.

I must also thank Professors Andrew Fear (University of Manchester), Christopher B. Zeichmann (University of Toronto), Javier Arce (Université Charles de Gaulle-Lille 3) and, again, Nicholas de Lange, for kindly accepting to read the first draft of this book. Their comments and timely corrections have been invaluable to me. Likewise, the thoughtful suggestions of Professor Domenico Vera (Università degli Studi di Parma) were of great value to me.

Moving on to Spain, I cannot forget the constant assistance I received from my colleague and friend, Sabino Perea Yébenes (UNED, Madrid). His critical spirit and his apt and timely suggestions are and have been a constant source of encouragement for me. Thank you, Sabino. The careful reading made by my other colleague-cum-friend, Fernando Bermejo Rubio (UNED, Madrid), has been no less invaluable. His astute and intelligent observations have "inevitably" enriched the end-result of this work. Thank you, Fernando.

I am also thankful to Mr. Michael Eric Moss for his excellent English translation and careful revision of the text of this book, as well as to Dr. David Soria Molina for his invaluable help with the prosopographic map.

Now that my work is finally over (or has it only begun?), I hope the results are worthy of the noble motives that gave rise to it and, especially, of the extraordinary people who inspired it …

… As always, my wife, Ana, has supported me and helped me—as no one else could—throughout this exciting adventure.

Figures

Introduction

In the recently published *Encyclopedia of the Roman Army* (2015), a specialist on the Roman army states that Jews were "sometimes enemies of Romans, always critical of Rome, never recruited as soldiers in the Roman army."[1] Were we to accept such an view, this book would be over before it was even written. However, I can only accept the first proposition ("sometimes enemies of Romans") as correct; others are not. Jews were not always critical of Rome (some favourable opinions are to be found, for instance, in the rabbinic literature)[2] and there is sufficient evidence of the presence and recruitment of Jews in the Roman army from the Late Republican period onwards. I am aware that, according to the old legal maxim, the *onus probandi* ("the burden of proof") is upon him who affirms (*affirmanti incumbit probatio*). Nevertheless, the forcefulness with which this specialist makes his negative statement, without any further explanations, and ignoring the rule of inference *modus tollendo tollens* (the very basis of falsification in the scientific method), is highly suspicious and casts serious doubt on the reasons that gave rise to it. In fact, it seems to be more of an inflexible reaction from a "purist" on the composition of the regular Roman army faced with the growing body of research that points to the presence of Jews in the Roman military.

Although there were moments of enormous tension between the Jewish people and the Roman State which led to bloody wars and harsh repression, we have information, from sources such as Josephus, Roman legislation, inscriptions, Egyptian papyri, Roman authors, Christian and rabbinic literature, about military service by Jews in the Roman army from the end of the Roman Republic (1st century BCE) to the time of Justinian (6th century CE).[3] Pieter Willem van der Horst, in his long list of jobs held by Jews according to epigraphic evidence, did not fail to include those in the military (centurion, army-reservist, military commander).[4] In fact, he writes "that this list contains a number of military functions may seem strange at first sight, but there is abundant evidence that did indeed serve in the armies of the Greek states and the Roman

1 Currant 2015, 559.

2 See de Lange 1978, 275–281 and 2017, 27–28. Cf. Lapin 2012.

3 According to Andrew J. Schoenfeld (2006, 117), "[…] with these facts in mind, the prospect of active Jewish participation in the military becomes tangible, and upon careful examination of primary sources it is possible to demonstrate the profound extent to which Roman Jews operated in the Imperial Army."

4 van der Horst 1991, 99.

 | DOI:10.1163/9789004507258_002

Empire."[5] Thus, as Jonathan Roth aptly pointed out, "Jews (in the religious if not the ethnic sense) served in the regular Roman army, and Jews in both senses fought side-by-side with Romans, and provided an important command function, both in peace and war. This role is generally ignored, even by books specifically detailing Roman, and pro-Roman forces."[6]

Nevertheless, it seems that there was also a reluctance to acknowledge this fact amongst certain conservative sectors of Jewish historiography. Authors such as Th. Reinach believed that Jews never served in the Roman army, either due to exemptions granted by Roman authorities or through payment of monetary compensation to the State (known as *aurum tironicum* in the latter Roman Empire).[7] There can be no doubt that the fact that Jews and Romans were opposed in at least two bloody wars and the theoretical incompatibility between Jewish military service and the normative exegesis (especially with regards to the Sabbath rest) found in rabbinic literature, which we nowadays consider to define "Jewish orthodoxy", hindered an in-depth analysis of the sources that revealed the presence of Jews in the Roman service. Some historians not only created a rigid and monolithic image of rabbinic Judaism, thus conditioning the definition of the very word "Jew" in ancient society, but they also projected certain Talmudic conventions (or even some from modern Judaism) onto previous periods.[8] The Jewish historiographical criticism that emerged from 19th century liberal circles certainly noticed the dynamic transmission of the central Jewish tradition and, according to Chanan Gafni, "[…] many additional doubts were raised with regard to the talmudic attributions, a part of a growing tendency to challenge the Talmud's credibility, and accordingly, its legitimacy as a reliable historical source. As a matter of fact, doubts were even raised regarding the attributions of statements with no parallels in the Talmud […]."[9]

Considering that the Roman Empire was responsible for the destruction of the Temple of Jerusalem, some religious scholars are still reluctant to acknowledge Jewish participation in the Roman military, "often labelling Jewish soldiers apostates or questioning their attachment to the Jewish community."[10] These historians assume that the unwavering religious requirements of "Jewish ortho-

5 van der Horst 2015, 3 note 27, and 52. Cf. Williams 1998b, 88–89, 97.

6 Roth 2019, 79.

7 Reinach 1900, 626.

8 Roth 2007, 409.

9 Gafni 2016, 80. See also Stern 1995; Avery-Peck and Neusner 1999, I, 123–230 ("The Debate about Talmudic History").

10 Schoenfeld 2006, 125 (= 2019, 106). While not denying the evidence for Jews serving in the Roman army, and even providing irrefutable proof that supported this reality, some

doxy" were an insurmountable obstacle for Roman military service. Nevertheless, we must remember that the religious practices and codes of behaviour of Jewish communities throughout the Roman Empire were so varied and changing that it is impossible to reduce them to a single model of "normative Judaism" from the texts of rabbinic scholarship.[11] Often the Rabbis make assumptions about the world around them, and these assumptions might be a starting point to draw inferences about historical conditions, but at the same time it seems that they envision an idealized world of the faithful.

Documents from the Jewish archives of Babatha and Salome Komaïse, discovered in the late 1950s in a village called Maoza on the south coast of the Dead Sea (within the Nabataean territory annexed by Rome in 106 CE as part of the new province of Arabia)[12] suggest the adoption of a broader consideration of the likely wide range of practices by Jews in the Greco-Roman world. The set of papyrus documents rolled up in a leather bag was found in a cave associated with Simeon bar Kokhba (the Cave of Letters at Naḥal Ḥever), the leader of the second Jewish revolt against Rome (ca. 132–135 CE). The archive of the Jewish woman Babatha of Maoza—who probably died during the revolt—contained documents of a legal nature (marriage contracts, guardianship arrangements, loans, land registrations, etc.).[13] These documents are written both in Greek and Aramaic as well as in the regional language of Nabatean, transacted in diverse legal jurisdictions, and involving Jews, Roman citizens (who may or may not have been Jews), and Nabateans.[14] Some of them, which were recorded with the public authorities, are not consistent with rabbinical rules. In fact, at the time at which they are dated—between 70 CE and the end of the Bar Kokhba revolt—Jewish law was in the process of being created in rabbinical schools. According to Kimberley Czajkowski, "it is worth pointing out again that 'Jewish' law and justice in this period in no way equate to rabbinic law or justice."[15]

Some documents from this archive contain an oath by the reigning emperor or by his τύχη (*týche*).[16] There is also another example of a Jew from the

---

authors, such as Shim'on Applebaum (1971, 184 note 16) believed that these were "renegades" or apostates.

11 Schoenfeld 2006, 125–126 (= 2019, 106), based on Barclay 1996, 416. On classical rabbinic Judaism, see Stemberger 2009.

12 Bowersock 1983, 2.

13 Wise 2015, 77–78.

14 Kraemer 2004, 143; Czajkowski 2017, 5–9.

15 Czajkowski 2017, 155. According to Hannah M. Cotton (1998, 172), these documents "are representative of Jewish society as a whole in the period under discussion. They present a faithful picture of the realities of life at the time that they were written".

16 For instance, Cotton 1995, 176, ll. 2–3: [῾Ο δεῖνα] Λειουου ὄμνυμι τύχην Κυρίου Καίσαρος κ[α]λῇ

Fayyum, Soteles son of Josepos, who swears by the emperor.[17] In this case, the editors of the *Corpus Papyrorum Judaicarum* believed that "an oath by the emperor presupposes his superhuman origin, which contradicts the principles of Judaism".[18] Thus, it was assumed that Jews at the time were less aware—even oblivious—of the religious implications of swearing by the emperor or by his *týche* (*genius*) from the viewpoint of a "monotheistic Jewish theology". However, it was very difficult to find justification in an alleged coercion—of which there is no evidence in the documents—[19] or in the assumption that they were Hellenised Jews, whose "strange behaviour" could somehow be excusable. Nevertheless, the Jews present both in the archive of Babatha and in that of Salome Komaïse, daughter of Levi, the sister of the declarant mentioned at the beginning of the legal text, sign and underwrite the documents in Aramaic, employing professional scribes whenever they needed their documents to be drafted in Greek.[20] The Jews represented in these papyri are from densely populated Jewish areas, from the heartland of the Bar Kokhba revolt, a religious and national movement. In fact, the presence of these documents in the caves demonstrates their owners' participation in the revolt: "the writers of these documents cannot and should not be regarded as assimilated Jews."[21] In this regard, it is worth keeping in mind that they describe themselves as Jews. In one of the preserved papyri (*P.Yadin* 12), Babatha's son Jesus appears explicitly described as a Ἰουδαῖος, a "Judaean"/"Jew",[22] and Babatha's own marriage certificate with her second husband Judah (*P.Yadin* 10) is often identified as an early example of a *ketubbah*—a Jewish marriage contract.[23]

πίστει ἀπογεγράφθαι ὡς προγέγραπται μηθὲν ὑποστειλάμενος ("I, X son of Levi, swear by the genius of the Lord Caesar that I have in good faith registered as written above, concealing nothing"); *P.Yadin* 16, ll. 33–35 (Lewis, Yadin and Greenfield 1989, 67). See Czajkowski 2015 on this oath. Cf. Belayche 2003 for a consideration of how *týche* was conceived in Roman. Palestine specifically.

17 CPJ II, 427 = BGU IV, 1068 (101 CE). See Czajkowski 2015, 189–190.

18 CPJ II, 214.

19 Czajkowski 2017, 32.

20 *P.Yadin* 11: Iustinus; *P.Yadin* 14, 15, 17, 18: Theenas son of Simon; *P.Yadin* 19: X son of Simon; *P.Yadin* 20–27: Germanus son of Joudah. See Cotton 1998, 168.

21 Cotton 1998, 173. Czajkowski 2017, 8: "Thus, while none of the principals describe themselves as Jews, there seems to be little doubt that we are dealing with a Jewish community here."

22 *P.Yadin* 12 (Lewis, Yadin and Greenfield 1989, 48–49).

23 *P.Yadin* 10 (Yadin, Cotton and Gross 2002, 118–127). See Czajkowski 2017, 8 and 204. According to this author, "Babatha's *ketubbah* may suggest some kind of awareness by the later rabbinic texts, based upon the inclusion of the mandatory *ketubbah* components that are listed in the Mishnah. These, however, can just easily demonstrate that rabbinic sages

The diversity and fluidity manifested in the documents from the Judaean desert are the evidence we have for the state of Jewish law and for the limited reach of the rabbis' authority at the time, which was still at an embryonic stage.[24] Nevertheless, the first rabbinic texts (*Mishnah, Tosefta, Jerusalem Talmud, Babylonian Talmud,* among others) present a misleading view of an early, rapid and uniform expansion of rabbinisim, projecting the notion of a consolidated and triumphant ideological current within Judaism—both in Palestine and in the Diaspora[25]—far further back in history than at the time when the rabbinical movement began to take shape. However, the absence of references to rabbis in the over six hundred inscriptions from Rome (dated between the 2nd and 5th centuries CE) has caused no shortage of problems for traditional historiography[26] when it came to reconciling epigraphic sources and rabbinic texts. Titles such as "archon", "gerousiarch", or "mother of the synagogue" hardly fit into rabbinic literature. Moreover, the laws of *Codex Theodosianus* (16.8.1 and 16.8.13) that referred to "archisynagogues", "principals", "elders" or "presbyters" were confusing to historians, as was the scarce presence of Hebrew and Aramaic during the first centuries of the common era,[27] and the striking absence of quotes from rabbinic literature, even from early texts such as the *Mishnah*.[28] In the 1920s, Ludwig Blau was the first scholar to draw attention to the differences between rabbinic literature and the archaeological and epigraphic record from both the Diaspora (Rome, North Africa) and from the Land of Israel itself (Capernaum and Chozarin).[29]

A few exceptions notwithstanding,[30] modern historiography is inclined to defend the notion of a gradual growth of the rabbinical current within the Judaism of the imperial era, which does not imply the immediate disappearance of other non-rabbinical Jewish expressions.[31] "My hypothesis—states

incorporated existing Near East or Jewish custom into their own law, rather than that Babatha and Judah were living under some kind of early rabbinic influence (direct or indirect) at this time." (2017, 155).

24 Cotton 1998, p. 172.

25 On the notion of "Diaspora", see Laham Cohen 2019.

26 For example Ginzberg 1909–1938; Krauss 1922; Foot Moore 1948–1954.

27 According to Harry J. Leon (1995, 226), "the inscriptions have clearly demonstrated that there was little, if any, knowledge of Hebrew among the Jews of imperial Rome, so that the opinions of the rabbis were probably quite unknown to them." Inscriptions fully in Hebrew and which, in some cases, featured Talmudic content would only appear in Venosa by the 9th century. See Colafemmina 2000.

28 See Laham Cohen 2021, 3.

29 Blau 1924 and 1926.

30 Collar 2013, 195 and *passim*. Cf. Edrei and Mendels 2010.

31 See Costa 2019. Cf. Friedheim, 2020, 312.

Martin Goodman—is that groups and philosophies known from pre-70 Judaism continued for years, perhaps centuries, after the destruction of the Temple."[32] When studying Jewish symbols in the Graeco-Roman era, Erwing Goodenough already explicitly defended that the rabbinical current was not the only Judaism that survived the destruction of the Second Temple in 70 CE.[33] In fact, some researchers believe that rabbinical Judaism was a marginal phenomenon, even in Palestine, until the 4th century CE[34] and that only then did it begin to gain visibility until it became the dominant current among Jewish communities throughout the Mediterranean.[35] In any event, it seems indisputable that the process of rabbinisation was a slow one[36] and that it coexisted antagonistically with other currents and groups that also claimed the heritage of Judaism. Even though the texts of these currents are now lost, the rabbis themselves sometimes refer to them.[37]

In fact, Jewish identity can be detected both in the "orthodox" environment of rabbinic schools and in a social context dominated by the cultural syncretism of Jews who, having adapted perfectly to a gentile environment, had adopted Greek and Latin names, and even pagan motifs in the decoration of their communities' synagogues or burial grounds.[38] Many of the Jews in surviving papyri and inscriptions, whose identity is clear from their names, epigraphic formulae or the use of known religious symbols, not only did not wish to hide their Jewishness but, in fact, appeared proud of it. As Margaret H. Williams points out, the epithet *Ioudaîos* in inscriptions is "used with pride by both those born as Jews and those converted to Judaism, it might, depending on place and time, serve either to stress similarities or to emphasise differences."[39] In this regard, it would be wrong to establish the degree of religious observance, as understood by the rabbinic authors of the Mishnah or Talmud, as the standard by which Jews must be identified.

The Sabbath, much more than Jewish dietary laws and conflicts arising because of oaths to emperors and other religious matters, seems to be the great-

32 Goodman 2004, 161.
33 Goodenough 1988, 25.
34 According to Lee Levine (2005, 472), rabbinical influence in the domain of the synagogues was very limited. See also Magness 2005.
35 Goodenough 1965, vol. 12, 198; Schwartz 2001; Schwartz 2002; Lapin 2012. Victor Rutgers dated the beginning of rabbinical influence in the Diaspora towards the 4th century (1998a, 27).
36 For example, Rutgers 1999.
37 On non-rabbinic Jewish groups, see, among others, Hayes 2007.
38 Figueras 2000–2002; Schoenfeld 2006, 117 (2019, 97); Weiss 2013; Friedheim 2015.
39 Williams 2013, 279 (= Williams 1997, 257). Cf. Schwartz 2013.

est stumbling block to regular Jewish service in foreign armies, but this was apparently overcome on a regular basis.[40] The Jews who served Roman army units seem very far removed from the world of the Talmud and the Rabbinic Judaism that was developing in the Roman World. To quote another of Andrew J. Schoenfeld's insightful observations, "Authors who question the authenticity of these soldiers' religious conviction detract from our understanding of Jewish history in the Roman period and render an injustice to the memory of these Roman Jews, who felt strong enough in their faith to participate in their religious community, make dedications in their synagogues and use Jewish markers on their graves."[41]

In broad outline, the main objective of the my research is to study the process leading from military exemption from service in the Late Roman Republic and the occasional presence of Jews in the Roman army—not always under specific privileged conditions—during the High Empire to their exclusion from the *militia armata* in the time of the Christian Empire (4th–6th centuries CE), not to mention the special situation recorded in the sphere of Roman Egypt and the creation of special military units entirely or chiefly made up of Jewish soldiers (such as the troopers of Zamaris and those under *dux* Syrianus). It is truly difficult to make a statement on the significance of the presence of Jews within the Roman army, either in specific military units or on an individual basis. It would be unwise to unhesitatingly accept the notion that, as Jean Juster argued, Jewish military service in the Empire had become commonplace.[42] It is worth keeping in mind that the mere acknowledgement of a reality is not grounds in itself for assigning it a greater level of importance than that which it deserves in all fairness. Nevertheless, accounts on the subject do not allow for a reduction of Jewish presence in the Roman army to a merely marginal phenomenon, nor can its minority nature be dismissed as an exceptional and insignificant episode.[43]

40 According to the ostraca preserved in Edfu (*Apollinopolis Magna*), in the south of Egypt, from the period between Vespasian and Trajan (69–117 CE), local Jews paid the special tax to the *fiscus Iudaicus* even on Saturdays.

41 Schoenfeld 2006, 126 (= 2019, 107).

42 Juster 1910, II, 265–276. Sevenster's interpretation (1975, 154–156) was much more cautious on this subject.

43 Roth 2007, 89.

CHAPTER 1

Jewish Military Service in Hellenistic Armies

1 Precedents

After the Assyrian conquest of the northern kingdom of Israel at the end of the eighth century BCE, the *Annals*, some inscriptions of Sargon II, and the Horse Lists found in Fort Shalmaneser (also called the Review Palace) in the modern city of Nimrud, reveal that entire Israelite-Samarian units—especially chariotry and cavalry ones—were integrated into the army of the new overlords.[1] According to certain authors, Jewish military service in foreign armies is likewise attested in the Hebrew Bible. If we are to accept P. Korgrun's interpretation of *Deuteronomy* 17:16 ("He [the king] shall not acquire many horses, nor, to add to his horses, shall he cause the people to go back to Egypt, for this is what the Lord said to you, 'You shall never go back that way'"), Haggai Olshanetsky believes that this Biblical passage hints to the existence of Jewish recruits to Pharaonic Egypt. According to this author, the Biblical restriction on the number of horses a Jewish king could own arose from a wish to avoid exchanging Jewish troops for Egyptian horses.[2] In fact, the author of the so-called *Letter of Aristeas* (§ 13), who displays a remarkable knowledge of the institutions and customs of the Ptolemaic court of Alexandria, states that, at the time of the last king of Judah, Zedekiah (597–587 BCE), there were Jewish soldiers serving in the Egyptian army of Psammetichus II (594–589 BCE) who waged war against the Ethiopians (Herodotus 2.61).[3] This military campaign is likewise confirmed by Greek inscriptions on the legs of the Rameses II's colossi, at the entrance of the temple of Abu Simbel.[4] In one of them it is stated that the Pharaoh's troops were made up from Greek, Egyptian and ἀλλόγλωσσοι ('those who speak other languages') troops, which almost certainly included Jewish mercenaries.[5]

This was the time of the foundation of the Jewish military colony of Elephantine (*Yeb*, from the Egyptian *'iebew*, 'elephant place'), an island on the Nile River

1 Dalley 1985; Dezsö 2012, vol. 2, 70–72; Olshanetsky 2015; Aster 2019, 605.

2 Olshanetsky (2015, 21) states that "this sentence provides us with evidence about Jewish military service in foreign armies during early antiquity, in exchange for material goods."

3 Wright III 2015, 138.

4 Bernard and Masson 1957.

5 See Juster 1914, II, 265 note 2; Porten 1968, 9–12; Méléze-Modrezejewski 1995, 24–25.

 | DOI:10.1163/9789004507258_003

at the southern border of Egypt. Although it has yielded a wealth of Aramaic papyri and *ostraka* that have survived to this day, we are entirely in the dark as regards its origins. Judging by the information contained in various documents dated 407 BCE, the Jews of Elephantine believed that the construction of their temple, dedicated to "Yahô", went back to "the day(s) of the kings of Egypt," that is, to a period prior to the conquest of Egypt by Cambyses in 525 BCE.[6] Thus, the establishment of this Jewish colony should be dated to the Saite period,[7] after which it went into the service of the Persian Empire.[8] There can be no doubt as to its military nature[9] which, owing to a decree issued by Darius, did not prevent the observance of Jewish religious prescriptions and festivities, such as the *Sabbath* or *Pesach* (Passover).[10] In spite of a number of discrepancies in religious practices, its relations with the motherland of Judaea are highlighted by a letter sent by Elephantine Jews to their brethren in Samaria and Jerusalem, requesting help to rebuild their temple dedicated to the God of Israel, which had been destroyed by the violent actions of a band of Egyptian rebels with the support of Vidranga (Waidrang), the governor of Syene.[11]

The soldiers of Elephantine served both as frontier guards and as a regional police force to assure law and order.[12] The papyri from the "Elephantine

6 Porten 2011, 143 (B19, l. 13), 148 (B20, l. 12), 150–151 (B21, ll. 4–5). See Kaplan, 2003, 8; Fischer-Bovet, 2014, 36–37. Jewish presence in Egypt, albeit in Eastern Delta, is attested in the 6th century by *Isaiah* 19, and *Jeremiah* 44:1 and 46:14. See Mahaffly 1895, 7 note 1; Schürer 1986, III.1, 39–41.

7 Joseph Mélèze-Modrezejewski (1995, 25–26) suggested that the origins of the Jewish colony of Elephantine may go back to the reign of Jehoiakim (609–598 BCE), "who chose to ignored the preachings of Jeremiah and remain faithful to Pharaoah Necho II, to whom he owed his kingship."

8 Porten 1968, 15–19.

9 According to Victor Tcherikover (1982, 334), "since the publication of the Elephantine papyri scholars can no longer question the military role of Jews in Egypt. The Jews of Elephantine call themselves 'the Jewish force' (חילא יהודיא), they live in 'the fortress' (בידחא) and are divided into 'standards' (דנלים), special battalions—and their entire life bears the stamp of military organization."

10 Porten 2011, 16, 281; *Ostracon Clermont-Ganneau*, No. 152 (ed. Dupont-Sommer 1949). See Schwartz 2017, 56. According to Gard Granerød (2019), "we probably get closer to a historically verifiable Persian-period Judean religion in the Elephantine documents than in the world of the thoroughly edited texts of the Hebrew Bible." On the religious life of the Jewish colony of Elephantine under Persian rule, see Joisten-Pruschke, 2008.

11 Mélèze-Modrezejewski 1995, 39; Eshel and Stone 2013, 95. Help was also requested from the highest Persian authorities: Porten 2011, 153 (B22, ll. 9–10), 151 (B21, ll. 9–11). See Fischer-Bovet 2014, 36.

12 Schwartz 2017, 55–59.

archives" (494–399 BCE) have revealed to us that the garrison (*hayla*) was divided into "companies" or "banners" (*degel*), the equivalent of the "thousands" of the old Israelite armies and, later on, of the Greek armies. Each company, which in turn was subdivided into "centuries," and these into "decuries," was led by a commander who lent it his name. The entire garrison was placed under the orders of a commander-in-chief (*rab hayla*) whose headquarters were in Syene (today known as Aswan); his title corresponded to that of a Greek *phroúrarchos*.[13]

We know that, under Persian authority, the Jewish soldiers who lived in Elephantine along with their families had regular pay. Some of them were also granted plots of land as a reward for their service. According to Mélèze-Modrezejewski, "the Achaemenids thus paved the way for the cleruchic system—concession of a parcel of land in return for military service—the Ptolemaic rulers were to establish after Alexander had conquered the country."[14]

The last document from the Elephantine archives is a letter written in 399 BCE by a Jew named Shawa, son of Zekhariah, to his friend Yislah in Memphis, which tells of Nepherites' accession to the throne.[15] The text reveals that the local temple had been rebuilt between the years 406 and 401 BCE, but the author of the letter also states that it was subsequently definitively destroyed. From that point onwards there are no more direct sources on the Jewish colony of Elephantine[16] or on a Jewish presence in Egypt until Alexander's conquest.[17] Nevertheless, it would be reasonable to assume that, after Achaemenid hegemony in Egypt was restored under Artaxerxes III (343 BCE), the system of Jewish military colonies according to the model in Elephantine would remain operational. Even though we lack direct information to confirm it, we know that the aforementioned king established a Jewish military settlement on the Caspian sea-coast in 340 BCE.[18] In any event, scarcely twelve years later, in 332/331 BCE, the Persians gave way before the unstoppable thrust of Alexander's armies, which led to a new political context that once again provides us with valuable

13 Mélèze-Modrezejewski 1995, 26.
14 Mélèze-Modrezejewski 1995, 26.
15 Kraeling 1953, No. 13; *TAD* A3.9. See Schürer 1986, III.1, 39–40.
16 Mélèze-Modrezejewski 1995, 43.
17 There is a papyrus of dubious provenance dated to the late 4th or early 3rd century BCE, written in Aramaic, with a list of payments featuring a few Jewish and Greek names (Cowley 1923, 190–199, No. 81; Grelot 1972, No. 13). See Schürer 1986, III.1, 40.
18 Olshanetsky 2015, 22.

sources of information on the socio-political situation of the Jews and their possible presence in the military.

Mainly relying on Hecataeus of Abdera (4th century BCE), Flavius Josephus states that Alexander's offer to respect their religious rules and customs (τοῖς πατρίοις ἔθεσιν) led Jews to enlist in his army:

> And, when he said to the people that if any wished to join his army while still adhering to the customs of their country, he was ready to take them, many eagerly accepted service with him.[19]

Even though the veracity of the tale of Alexander's trip to Jerusalem, in which this story takes place, has been challenged, the fact remains that other, less suspicious passages of Josephus' works describe the presence of Jewish soldiers among his troops. We may therefore assume that, regardless of the author's apologetic intent, the recruitment of Jewish soldiers on the above terms contains a grain of historical truth.[20] Relying on Hecataeus once again, Josephus states that Alexander severely punished Jewish soldiers in his army for refusing to participate in the rebuilding of the temple of Bel in Babylon,[21] although he exempted them from that task shortly afterwards when he understood that it was incompatible with their religious views.[22] The next anecdote, which took place during the march towards the Erythraean Sea, tells of a Jewish archer, Meshullam, who shot a bird pagan soldiers had released in an attempt at divination,[23] highlighting the latent tension between Jews and gentiles in an army made up from such varied contingents. In any event it can be said that, as happened with other peoples in their surroundings (such as the Ituraeans, at a later date[24]), the Jews also stood out among Alexander's troops as experienced archers,[25] and would remain as such in the various armies in which they served.[26]

---

19 *Ant. Iud.*, XI, 339: εἰπόντος δ' αὐτοῦ πρὸς τὸ πλῆθος, εἴ τινες αὐτῷ βούλονται συστρατεύειν τοῖς πατρίοις ἔθεσιν ἐμμένοντες καὶ κατὰ ταῦτα ζῶντες, ἑτοίμως ἔχειν ἐπάγεσθαι, πολλοὶ τὴν σὺν αὐτῷ στρατείαν ἠγάπησαν.

20 Mahaffly 1895, 86; Juster 1914, II, 266 note 4; Launey 1987, I, 550–551.

21 This is also attested by Arrian, *Anab. Alex.*, VII, 17, 1–3 and Strabo, XVII, 1, 5. See Batsch 2005, 253. On the Jewish refusal to rebuild pagan temples, see *Midrash Aboda Zara* I, 7.

22 Josephus, *Contr. Ap.*, I, 22, 192–193. See Juster 1914, II, 265–266 note 4.

23 Josephus, *Contr. Ap.*, I, 22, 200–201.

24 Myers 2010, 41.

25 Gichon and Vitale 1991, 252: "The East was archer country par excellence since very early times, for the Jews no less than for other peoples."

26 Josephus, *Ant. Iud.*, XIII, 135; XVII, 23–25. See Gichon and Vitale 1991, 252–253.

It is obvious that Alexander also recruited Jewish mercenaries to bolster his forces.[27] According to Josephus, despite the tension that frequently arose with the Greeks, Alexander allowed Jewish settlers to live in the new city he had founded on the Nile Delta in acknowledgement of their outstanding contribution (συμμαχία) to the conquest of Egypt:

> At Alexandria there had been incessant strife between the native inhabitants and the Jewish settlers since the time when Alexander, having received from the Jews very active support against the Egyptians, granted them, as a reward for their assistance, permission to reside in the city on terms of equality with the Greeks.[28]

According to the text, it seems certain that, under Alexander, new Jewish military colonies were established in Egypt. The fact that Diodorus Siculus reports that circumcision was practiced there seems to confirm their existence in the times of Alexander,[29] and this reference has allowed Simon Claude Mimouni to hypothesize that the Jewish settlers had reached Egypt with Alexander's armies.[30] Nevertheless, it is also possible that these Jews descended from former Persian military colonies and were given permission to live in a newly-founded city in the Delta.[31] In any event, we are told that, following this same model, Alexander planned to send Samaritan soldiers to Upper Egypt, where "he would give them allotments of land, as in fact he did shortly afterwards, in the Thebaid, and this territory he ordered them to guard."[32]

---

27 Launey 1987, I, 542. Cf. Olshanetsky 2018, 8–9.

28 Josephus, *Bell. Iud.*, II, 487: Κατὰ δὲ τὴν Ἀλεξάνδρειαν ἀεὶ μὲν ἦν στάσις πρὸς τὸ Ἰουδαϊκὸν τοῖς ἐπιχωρίοις ἀφ' οὗ χρησάμενος προθυμοτάτοις κατὰ τῶν Αἰγυπτίων Ἰουδαίοις Ἀλέξανδρος γέρας τῆς συμμαχίας ἔδωκεν τὸ μετοικεῖν κατὰ τὴν πόλιν ἐξ ἴσου † μοίρας πρὸς τοὺς Ἕλληνας. Cf. Josephus, *Contr. Ap.*, II, 4, 42–44. See Schürer 1986, III.1, 42–43.

29 Diodorus of Sicily, *Bibliotheca historica*, I, 55, 5: ὅτι δὲ τοῦτο τὸ γένος Αἰγυπτιακόν ἐστι σημεῖον εἶναι τὸ περιτέμνεσθαι τοὺς ἀνθρώπους παραπλησίως τοῖς κατ' Αἴγυπτον, διαμένοντος τοῦ νομίμου παρὰ τοῖς ἀποίκοις, καθάπερ καὶ παρὰ τοῖς Ἰουδαίοις ("And the proof which they offer of the Egyptian origin of this nation is the fact that the Colchi practise circumcision even as the Egyptians do, the custom continuing among the colonists sent out from Egypt as it also did in the case of the Jews").

30 Mimouni 2007, 294–295.

31 On Jewish presence at the time of the founding of Alexandria, see Kasher 1985, 189; Empereur 2018, 11.

32 Josephus, *Ant. Iud.*, XI, 345: ἐκεῖ γὰρ αὐτοῖς δώσειν κλήρους γῆς· ὃ καὶ μετ' ὀλίγον ἐποίησεν ἐν τῇ Θηβαΐδι φρουρεῖν τὴν χώραν αὐτοῖς προστάξας.

2 Under the Ptolemies

As Jean Lesquier showed in his classic study on Ptolemaic military organization, the three basic pillars of their army were military settlers (*kátoikoi*), mercenaries, and indigenous inhabitants.[33] Sources reveal a strong presence of Jews in each of these spheres (according to this scholar's estimates, they made up 20% of all Ptolemaic forces), even though the first category makes up the bulk of the most abundant and detailed information available to us, even with regards to the very beginning of the Lagid dynasty. On the subject of the aforementioned *Letter of Aristeas*, the author points out that, once the city of Jerusalem was taken, Ptolemy Lagus "deported one hundred thousand from the land of the Judaeans into Egypt, from whom he armed about thirty thousand chosen men and settled them in garrisons in the country."[34] Even if we are to acknowledge that, as is common in ancient sources, these figures may be over-estimated,[35] the creation of Jewish military colonies for defensive purposes was a historical fact.[36] Josephus confirms the information provided by the author of the *Letter of Aristeas* with official documents from the Ptolemaic court, which Apion, his antagonist, was able to check.[37] The new regime's distrust towards the majority of the Egyptian population, which vastly outnumbered them, was undoubtedly a factor that led to the inclusion of Jews and other ethnic components in their armies[38] and the adoption of the old Per-

33 Lesquier 1911, 11–29. Cf. Bar-Kokhva 1979, 22–26.

34 *Letter of Aristeas*, 12–13: [...] ἐν ὅσῳ καὶ πρὸς δέκα μυριάδας ἐκ τῆς τῶν Ἰουδαίων χώρας εἰς Αἴγυπτον μετήγαγεν, ἀφ' ὧν ὡσεὶ τρεῖς μυρίαδας καθοπλίσας ἀνδρῶν ἐκλεκτῶν εἰς τὴν χώραν κατῴκισεν ἐν τοῖς φρουρίοις [...]. According to Josephus, Ptolemy I also sent many Jews to Cyrenaica and Libya to strengthen his grip on those regions (*Contr. Ap.*, II, 44). See Hengel 1989, 50–51.

35 Anagnostou-Canas 1994, 174 note 117; Barclay 1996, 22; Collins 2000, 66; Fischer-Bovet 2014, 54.

36 Mélèze-Modrezejewski 1995, 83. According to Seth Schwartz (2017, 61), "much of the Jewish settlement in rural Hellenistic Egypt was military."

37 Josephus, *Contr. Ap.*, II, 62. For instance, the sack of Jerusalem by Ptolemy I is confirmed by Agatharchides of Cnidus (quoted by Josephus himself, *Contr. Ap.*, I, 209–211; *Ant. Iud.*, XII, 4–6) and by Apianus, *Syr.*, 50 (252). See Schürer 1986, III.1, 41; Juster 1914, II, 266 note 2; Launey 1987, I, 543.

38 A necropolis in Alexandria has yielded a number of tombs that belonged to individuals of various nationalities and which have been attributed to mercenaries from the early Ptolemaic period (early 3rd century BCE) who were buried there. One of the epitaphs features the name of one 'Aqabiah son of Elyōenai' (*CIJ* II, No. 1424 = *JIGRE*, No. 3). See Schürer 1986, III.1, 47; Tcherikover 1982, 335; Kasher 1985, 189. In fact, based on epigraphy, the Jews of Alexandria were the third largest group of the people living in the city (Savvopoulos 2018, 126).

sian system of military settlements, especially on their borders.[39] In the letter he allegedly sent to High Priest Eleazar, king Ptolemy II Philadelphus (288–247 BCE) stated that he had followed his father's example by employing Jews "in their prime" in his army and even raising some of them to high military positions.[40] In fact, he also tells us of the founding of Jewish garrisons under his predecessor:

> Since it has happened that a large number of Judaeans have settled in our country, whom the Persians took from Jerusalem during the time when they held power and yet who also came with our father into Egypt as prisoners of these he assigned a good number to the regular military at higher than mercenary pay, and similarly also those who were already there whom he judged faithful, to them he gave garrisons that he built so that native Egyptians might [not] be afraid of them, but now we, having inherited the kingdom, meet all people in a more benevolent manner, but more so especially your fellow citizens.[41]

Indeed, papyrological information bears witness to the fact that Jewish military colonies were already fully consolidated by the time of Ptolemy II Philadelphus (288–247) and Ptolemy III Euergetes (247–181).[42] *Kátoikoi* were active settlers-soldiers of alien nationality whose *klēroi* were passed down as an inheritance from father to son.[43] The descendants of these soldiers, known as *Epígonoi*, were a constant source of renewal for the contingents which, in practice, constituted a veritable reserve army.[44] However, sources also display clear evidence

---

39 Feldman 1993, 87.

40 *Letter of Aristeas*, 37: [...] εἴς τε τὸ στράτευμα τοὺς ἀκμαιοτάτους ταῖς ἡλικίαις τετάχαμεν, τοὺς δὲ δυναμένους καὶ περὶ ἡμᾶς εἶναι, τῆς περὶ τὴν αὐλὴν πίστεως ἀξίους, ἐπὶ χρειῶν καθεστάκαμεν [...] ("[...] We have appointed to the army those who are at the prime of military age, and we have established over business matters those who are capable enough to attend us, who are worthy of trust by the court").

41 *Letter of Aristeas*, 35–36: [...] ἀφ' ὧν πλείονας εἰς τὸ στρατιωτικὸν σύνταγμα κατεχώρισεν ἐπὶ μείζοσι μισθοφορίαις, ὁμοίως δὲ καὶ τοὺς προόντας κρίνας πιστοὺς φρούρια κτίσας ἀπέδωκεν αὐτοῖς, ὅπως τὸ τῶν Αἰγυπτίων ἔθνος φόβος [μὴ] ἔχῃ διὰ τούτων· καὶ ἡμεῖς δὲ παραλαβόντες τὴν βασιλείαν φιλανθρωπότερον ἀπαντῶμεν τοῖς πᾶσι, πολὺ δὲ μᾶλλον τοῖς σοῖς πολίταις.

42 *CPJ* I, No. 18 (260 BCE), No. 19 (226 BCE), No. 22 (201 BCE). See Mahaffly 1895, 183; Schürer 1986, III.1, 42–43, 50–52; Launey 1987, I, 545; Mélèze-Modrezejewski 1995, 83. According to Lester L. Grabbe (2008, 194), "a good portion of the Jews in early Ptolemaic period seem to have been members of military units."

43 Schürer 1986, III.1, 42; Tcherikover 1982, 334–335; Kasher 1985, 3, 39–47; Mélèze-Modrezejewski 1995, 83; Grabbe 2008, 167, 196 and 219 ff.

44 *CIJ* I, 19, 20, 21, 23, 24, 26. See Lesquier 1911, 30 ff.; Schürer 1986, III.1, 52; Tcherikover 1982,

of Jewish military service in the regular Ptolemaic army, both in cavalry and infantry companies. Thus, for instance, a papyrus fragment related to a deed of renunciation (συγγραφὴ ἀποστασίου) that took place in Phebichis (Herakleopolites) in 260 BCE mentions a Jew named Alexander, son of Andronikos, a cavalry trooper (τῶν Ζωίλου δεκανικοῦ) on active service,[45] and Agathokles, son of Ptolemy, received a special identification as "Jew of the infantry battalion of Molossos (the name of the commanding officer, *taktómisthos*) stationed in the Herakleopolite nome."[46] In the same papyrus there appear two Jewish cavalrymen "of the first hyparchy." Documents such as this allow us to deduce that the Jews served in mixed units (hyparchies) of the regular army[47] alongside soldiers from "pseudo-ethnic groups," such as "Persians" or "Macedonians", grouped into *políteumata* or associations of soldiers with a common ethnos.[48] In the colony of Toubias in Transjordania, one of the *cleruchs* bore the name of "(..........) son of Ḥananiah, a Persian."[49] Considering that this is a Hebrew name, it is clear that this *cleruch* was not a Persian by race, but a Jew serving in a battalion of "Persians."[50] In fact, sources do not record a single exclusively Jewish unit until the time of Onias IV (i.e. the middle of the second century BCE).[51] Conversely, according to Victor A. Tcherikover, "we see quite a large number of Jews serving in the various branches of the Ptolemaic forces equally with the other Greek and 'barbarian' peoples, in harmony with the practice of the Hellenistic period of entrusting the defence of the state to formations of foreign mercenaries."[52] This reveals a high level of cultural assimilation among Jewish soldiers[53]

335; Feldman 1993, 87; Collins 2000, 116. On the legal status of Jews during their military service, see Kasher 1985, 48–55.

45 *CPJ I*, No. 18 (April, 260 BCE). See Juster 1914, II, 267 note 2.

46 *CPJ I*, No. 24 (16th April, 174 BCE). See Tcherikover 1982, 517 note 8; Schwartz 2017, 63.

47 *CPJ I*, 28.

48 Launey 1987, II, 1031; Honigman 2003. Although these Ptolemaic *políteumata* seem to have originated in a clearly military context, some of the soldiers' associations (*synodoi*) did not adopt this name, and, over time, the term entered widespread use beyond the realm of soldiering. See Lüderitz 1994, 200–203; Grabbe 2008, 182, Fischer-Bovet 2014, 291 and Sänger 2019, 96–100. A specific study of the Jewish *politeuma* of mid-2nd century BCE Heracleopolis in Cowey and Maresch 2001, and Sänger 2019, 33–37, 123–140. An opposite view in Ritter 2011.

49 *CPJ I*, No. 1 (April–May, 259 BCE).

50 Tcherikover 1982, 517 note 9; Rocca 2010, 28.

51 Even if we were to acknowledge that a purely Jewish unit is not mentioned in documents, and certainly did not exist until the days of Onias IV, Aryeh Kasher (1978) defended the notion that the evidence points towards the fact that there also were Jews organized in separate units under Jewish command even before the days of Onias IV.

52 Tcherikover 1982, 335.

53 According to Joseph Mélèze-Modrzejewski (1995, 87), "the sole factor distinguishing the

which is especially reflected in their names. The aforementioned researcher found that only about a quarter of the names in papyri relating to soldiers and military settlers in third- and second-century Egypt were Hebrew, the rest being Greek.[54] From the list of names of ten Jews featured on a late-3rd century BCE papyrus, nine of them were Greek names and only one was Hebrew.[55] Jewish names in Egypt fall into six basic categories: Hebrew names, Greek names phonetically close to a Hebrew name, Greek translations of Hebrew names, names with the theophoric element *théos* 'God', dynastic names, and names that are purely Greek with no specific connection to the bearer's Jewish identity.[56] In this regard, we must consider that the fact that many assimilated Jews called themselves by Greek names in documents (inscriptions, papyri, *ostraca*) does not allow us to distinguish them unless the Jewish identity of the bearer is specifically indicated.[57]

The full integration of Jews in the staff of the Ptolemaic army is also clear from the fact that some managed to enter the officer corps, or even rose to the highest military ranks.[58] A Jewish *taktómisthos* (τακτόμισθος) is mentioned in *CPJ I*, 24 and, as *CPJ I*, 28 has shown, Jews were part of the officers' complement at Samareia. A cavalry officer is the ἐπιστάτης ἱππαρχίας in *CPJ I*, 27.[59] One of the first high-ranking Jewish officers was Toubias, commander of the mentioned *cleruchy* (military settlement) in Transjordania whose men came from a variety of ethnic backgrounds.[60] A late-2nd century BCE inscription from Fayyum features the name of the ἡγεμών Eleazar, son of Nikolaos.[61] However, it is Flavius Josephus who provides the greatest wealth of information on those Jewish figures who reached the highest ranks in the Ptolemaic army. According to this historian, in the context of the dispute that had emerged early in the reign of Antiochus Epiphanes in 175 BCE (the Ptolemies had lost control over the land

Jew from others is his abiding attachment to the faith of his ancestors, the mark of his difference."

54 *CPJ I*, 27–28.

55 *CPJ I*, No. 22 (12th August, 201 BCE). According to Victor A. Tcherikover, "The enormous preponderance of Greek over Hebrew names may be illustrated especially by the names of Jewish soldiers and military settlers in the Fayûm in the third and second century B.C." (*CPJ* I, 27).

56 *CPJ I*, XVII–XIX, 27–30; Launey 1987, I, 550; Clarysse 1994, 199–200; Mélèze-Modrezejewski 1995, 117; Grabbe 2008, 145.

57 Tcherikover 1982, 523 note 3.

58 Launey 1987, I, 546–547; Ritter 2015, 117 and 120.

59 *CPJ I*, 147.

60 *CPJ I*, No. 1 (April–May, 259 BCE), No. 4 (12th May, 257 BCE), No. 5 (12th May, 257 BCE). See Launey 1987, I, 544; Kasher 1985, 46.

61 *JIGRE*, No. 115 = *CIJ II*, No. 1531. See Launey 1987, I, 545.

of Israel in 198 BCE) over the succession of the high priesthood in Jerusalem, Onias IV, who was apparently the legitimate heir, was forced to seek refuge in Egypt.[62] Despite his lack of previous military experience, King Ptolemy VI Philometor (181–145 BCE) entrusted him, along with one Dositheos, with the supreme command of all his armies while giving him permission to erect a Jewish temple in Leontopolis (Tell el-Yehudieh), in the Heliopolite Nome (on the eastern shore of the Delta).[63] It is very likely that this temple was associated to a Jewish military settlement known as ἡ Ὀνίου χώρα, 'the land of Onias'.[64] In fact, the Jews who served there must have made up a considerable force, as they would later display enough military capacity to prevent the passage of an army marching from Palestine to the support of Caesar.[65]

After the death of Ptolemy VI, Onias can be seen at the head of the armies of the widowed queen, Cleopatra II, and of her son and successor Ptolemy VII Philopator, waging war against the deceased king's brother, Ptolemy VIII Physcon Euergetes II, who was, in turn, supported by the Roman troops of L. Minucius Thermus. Despite the fact that, once he reached power, Ptolemy VIII initially persecuted the Jews for having fought on the wrong side, they were soon reconciled and this allowed for a new migration of Jews to Egypt.[66] In any event, after his death, his widow Cleopatra III (daughter of Ptolemy VI and Cleopatra II), who reigned as regent for her son Ptolemy X Alexander, benefitted from the military experience of two of Onias' children, Chelkias and Ananias, when she appointed them generals of her armies ca. 105–104 BCE in the war against her son Ptolemy IX Lathyrus.[67] Josephus provides the following description based on a lost passage from the works of Strabo:

62 Flavius Josephus mentions the existence of close bonds between the circle of power associated to Onias and his descendants and the Ptolemaic court. See Tuval 2013, 84.

63 Josephus, *Contr. Ap.*, II, 49–50; *Ant. Iud.*, XII, 387–388; XIII, 62–73; *Bell. Iud.*, I, 33; VII, 426–432. See Launey 1987, I, 546–547; Hegermann 1989, 141–142; Mélèze-Modrezejewski 1995, 117; Collins 2000, 69–78; Capponi 2007; Pearce 2017, 49–50.

64 Josephus, *Ant. Iud.*, XIV, 131; *Bell. Iud.*, I, 190. *Cf. JIGRE*, No. 38 = *CIJ* II, No. 1530: Ὀνίου γᾶ τρυφὸς ἁμετέρα in a metrical epitaph. See Tcherikover 1982, 283; Kasher 1985, 8; Mélèze-Modrezejewski 1995, 117; Barclay 1996b, 29; Collins 2000, 71.

65 Josephus, *Ant. Iud.*, XIV, 131; *Bell. Iud.*, I, 190–192. See Schürer 1986, III.1, 48.

66 Josephus, *Contr. Ap.*, 2, 50–57.

67 Launey 1987, I, 548; Goldstein 1989, 337; Olshanetsky 2015, 24; Pearce 2017, 50. The son of one Chelkias (not Chelkias himself) is mentioned as a *strategós* and honoured with a golden crown in a fragment of an inscription from an uncertain date unearthed in the Heliopolite nome (*CIJ* II, No. 1450 = *CPJ* III, No. 1450 = *JIGRE*, No. 129). See Schürer 1986, III.1, 48; Juster 1914, II, 267 note 2.

> […] For Queen Cleopatra, who was at war with her son Ptolemy, surnamed Lathyrus, appointed as her generals Chelkias and Ananias, sons of the Onias who had built the temple in the nome of Heliopolis, which was similar to the one at Jerusalem, as we have related before. And having entrusted her army to them, Cleopatra did nothing without their approval, as Strabo of Cappadocia also testifies, when he writes as follows. "For the majority, both those who came back from exile and those who were later sent to Cyprus by Cleopatra, immediately went over to Ptolemy. And only the Jews of the district named for Onias remained faithful to her, because their fellow-citizens Chelkias and Ananias were held in special favour by the queen."[68]

This text may allow us to conclude that the Jewish military unit from Onias' territory (known as Onias' *kátoikoi*), which was led by these commanders, was separate from the Ptolemaic army.[69] It seems that, when the war moved from Cyprus to Coele-Syria and Palestine in 104–102 BCE, the queen herself accompanied her army, which remained under the command of Chelkias and Ananias.[70] Her allies included the young Hasmonean prince Alexander Jannaeus, who had recently ascended to the throne, and whose kingdom had been singled out as a possible target for invasion by some of Cleopatra III's counselors. Nevertheless, Ananias (for Chelkias had died in Coele-Syria[71]) managed to dissuade the queen from pursuing this plan by stating that such an invasion would not only be unjust, but it would also earn her the enmity of all Egyptian Jews: "an injustice done to this man will make all us Jews your enemies."[72] As Josephus points out, Ananias' advice was grounded in the deep feeling of closeness Egyptian Jews felt towards their Palestinian brethren and towards the Judaism they themselves professed. In this regard, we must not forget that, roughly fifty

---

68 Josephus, *Ant. Iud.*, XIII, 285–287: […] Κλεοπάτρα γὰρ ἡ βασίλισσα πρὸς τὸν υἱὸν στασιάζουσα Πτολεμαῖον τὸν Λάθουρον ἐπιλεγόμενον κατέστησεν ἡγεμόνας Χελκίαν καὶ Ἀνανίαν υἱοὺς ὄντας Ὀνίου τοῦ οἰκοδομήσαντος τὸν ναὸν ἐν τῷ Ἡλιοπολίτῃ νομῷ πρὸς τὸν ἐν τοῖς Ἱεροσολύμοις, ὡς καὶ πρόσθεν δεδηλώκαμεν. παραδοῦσα δὲ τούτοις ἡ Κλεοπάτρα τὴν στρατιὰν οὐδὲν δίχα τῆς τούτων γνώμης ἔπραττεν, ὡς μαρτυρεῖ καὶ Στράβων ἡμῖν ὁ Καππάδοξ λέγων οὕτως· "οἱ γὰρ πλείους, οἵ τε συνελθόντες καὶ οἱ ὕστερον ἐπιπεμπόμενοι παρὰ τῆς Κλεοπάτρας εἰς Κύπρον, μετεβάλοντο παραχρῆμα πρὸς τὸν Πτολεμαῖον· μόνοι δὲ οἱ ἐκ τῆς Ὀνίου γενόμενοι Ἰουδαῖοι συνέμενον διὰ τὸ τοὺς πολίτας αὐτῶν εὐδοκιμεῖν μάλιστα παρὰ τῇ βασιλίσσῃ Χελκίαν τε καὶ Ἀνανίαν." Cf. Josephus, *Ant. Iud.*, XIII, 348–355. See Juster 1914, II, 267 note 2.

69 Tcherikover 1982, 283.

70 Josephus, *Ant. Iud.*, XIII, 349.

71 Josephus, *Ant. Iud.*, XIII, 351.

72 Josephus, *Ant. Iud.*, XIII, 354: "[…] εἰ τὸ πρὸς τοῦτον ἄδικον ἐχθροὺς ἅπαντας ἡμᾶς σοι τοὺς Ἰουδαίους κατασκευάζει." See Tcherikover 1982, 283; Bar-Kochva 2010, 93; Tuval 2013, 84.

years later (55 BCE), the Jewish soldiers in the Ptolemaic army who were guarding the frontier near Pelusium granted the troops of the Roman proconsul of Syria, Aulus Gabinius, free passage into Egypt, simply because the Jewish head of state, Antipater—the father of the future Herod the Great—, whom they respected, had asked them to do so.[73] A very similar event took place in 48 BCE, when the Jewish border garrisons were persuaded by a letter from High Priest Hyrcanus to grant free passage to an allied military force travelling to Egypt to assist Julius Caesar.[74] Once again, this behaviour highlights the close bonds between Egyptian Jews and Judaea.[75]

3 Under the Seleucids

Similarly to what took place in Egypt under the Ptolemaic dynasty, many Jews served in the Seleucid army from the outset and, alongside other ethnic groups, they participated in the same system of military colonies that was implemented for defensive purposes, especially in Seleucid-ruled territory in Asia Minor.[76] Seleucus I Nicator (312–281 BCE), the founder of the dynasty, granted the Jews citizenship rights and other honours for their invaluable assistance in his military campaigns:

> They also received honour from the kings of Asia when they served with them in war. For example, Seleucus Nicator granted them citizenship in the cities which he founded in Asia and Lower Syria and in his capital, Antioch, itself, and declared them to have equal privileges with the Macedonians and Greeks who were settled in these cities, so that this citizenship of theirs remains to this very day […].[77]

73 Josephus, *Ant. Iud.*, XIV, 99. See Mahaffy 1895, 441; Tcherikover 1982, 284; Shatzman 1991, 133; Seeman 2013, 302.

74 Josephus, *Ant. Iud.*, XIV, 131–132. See Mahaffy 1895, 458; Launey 1987, I, 549; Hegermann 1989, 145; Feldman 1993, 89 and 420; Olshanetsky 2018, 14–15.

75 Kasher 1985, 11; Bar-Kochva 1996, 54–142 and 182–252; Feldman 1993, 50–51 and 89–90.

76 Launey 1987, I, 551. According to Bezalel Bar-Kochva (1979, 45), "the troops of Asia Minor were the most diversified: in the Lydian and Phrygian *katoikíai* Greeks, Mysians, Jews, and Persians are definitely traceable among the settlers, and the Macedonians formed only a tiny minority." See also Grabbe 2008, 173; Roux 2020, 17.

77 Josephus, *Ant. Iud.*, XII, 199: "Ετυχον δὲ καὶ τῆς παρὰ τῶν βασιλέων τῆς Ἀσίας τιμῆς, ἐπειδὴ συνεστράτευσαν αὐτοῖς καὶ γὰρ Σέλευκος ὁ Νικάτωρ ἐν αἷς ἔκτισεν πόλεσιν ἐν τῇ Ἀσίᾳ καὶ τῇ κάτω Συρίᾳ καὶ ἐν αὐτῇ τῇ μητροπόλει Ἀντιοχείᾳ πολιτείας αὐτοὺς ἠξίωσεν καὶ τοῖς ἐνοικισθεῖσιν ἰσοτίμους ἀπέφηνεν Μακεδόσιν καὶ Ἕλλησιν, ὡς τὴν πολιτείαν ταύτην ἔτι καὶ νῦν διαμένειν […]. Cf. Josephus, *Contr. Ap.*, II, 39.

Indeed, it seems that, from that point onwards, unlike Syrians and Babylonians, the presence of Jews in the Seleucid army became commonplace.[78] Nevertheless, it is quite possible that Josephus' reference to the acknowledgement of citizenship and other privileges to Jews in the cities in Asia and Syria founded by the first Seleucid king may have implied the establishment of military colonies in those places. In any event, the letter sent by Antiochus III to his governor Zeuxis in 205 BCE reveals that the Seleucid king ordered the displacement of 2,000 Jewish families from Mesopotamia to the most important places in Lydia and Phrygia as military colonists to help him secure his hold over this region.[79] Josephus reproduces (or recreates) the contents of the letter while explaining the manner in which military colonies functioned:[80]

> [...] Learning that the people in Lydia and Phrygia are revolting, I have come to consider this as requiring very serious attention on my part, and, on taking counsel with my friends as to what should be done, I determined to transport two thousand Jewish families with their effects from Mesopotamia and Babylonia to the fortresses and most important places. For I am convinced that they will be loyal guardians of our interests because of their piety to God, and I know that they have had the testimony of my forefathers to their good faith and eagerness to do as they are asked. It is my will, therefore—though it may be a troublesome matter—that they should be transported and, since I have promised it, use their own laws. And when you have brought them to the places mentioned, you shall give each of them a place to build a house and land to cultivate and plant with vines, and shall exempt them from payment of taxes on the produce of the soil for ten years [...].[81]

---

78 Bar-Kochva 1979, 52. According to Olshanetsky (2015, 22), "despite of hostility and rivalry, collaboration between Jews and Seleucids never ceased. Recruitment of Jews, probably most of them *Mityavnim*, was a common feature." A Jewish source (*2 Maccabees* 8:20) mentions the participation of 8,000 Jewish soldiers and 4,000 Macedonians in the army raised by the Seleucids to repel the Galatian invasion that may have taken place during the so-called "War of the Brothers" in 229/228 BCE. See Bar-Kochva 1989, 500–507; Sekunda 1994, 19.

79 See Juster 1914, II, 268 note 1; Launey 1987, I, 551; Hengel 1989, 70; Van der Horst 2014, 145; Nodet 2019, 63–64; Roux 2020, 3–5. According to Nick Sekunda (1994, 13), "the community of Jewish ex-mercenaries who were settled in Phrygia and Lydia by Antiochus III appears nowhere in battle-lists, and so seems never to have been mobilized."

80 Schalit 1960.

81 Josephus, *Ant. Iud.*, XII, 147–151: [...] εἰ ἔρρωσαι, εὖ ἂν ἔχοι, ὑγιαίνω δὲ καὶ αὐτός. πυνθανόμενος τοὺς ἐν Λυδίᾳ καὶ Φρυγίᾳ νεωτερίζοντας μεγάλης ἐπιστροφῆς ἡγησάμην τοῦτό μοι δεῖσθαι, καὶ βουλευσαμένῳ μοι μετὰ τῶν φίλων, τί δεῖ ποιεῖν, ἔδοξεν εἰς τὰ φρούρια καὶ τοὺς ἀναγκαιοτάτους

The first book of Maccabees tells us of the presence of Jews in the army of Demetrius I Soter (161–150 BCE) which was sent to Judaea under the command of Greek general Bacchides. When the Seleucid king felt threatened by Alexander Balas (150–145 BCE), he attempted to draw High Priest Jonathan to his cause by assuring him that the religious customs of troops recruited among the Jews would be respected:

> Jews shall be enlisted in the forces of the crown to the number of thirty thousand men; they shall receive the standard rate of army pay. Some of them shall be stationed in the important royal strongholds, others placed in positions of trust in the kingdom. Their commanders and officers shall be of their own race, and they may follow their own customs, just as the king has ordered for Judaea.[82]

It is worth noting that, even during the Maccabean revolt, there were Jews in the Seleucid ranks fighting against their fellow believers.[83] Not even Josephus, who was an admirer of the Maccabees and the Hasmonean dynasty, could hide this fact. Moreover, he acknowledges that *Mityavnim* (Jews who embraced Hellenism and were ready to abandon old Jewish strictures) were not the only Jewish recruits to the Seleucid army at that period.[84]

τόπους τῶν ἀπὸ τῆς Μεσοποταμίας καὶ Βαβυλωνίας Ἰουδαίων οἴκους δισχιλίους σὺν ἐπισκευῇ μεταγαγεῖν. πέπεισμαι γὰρ εὔνους αὐτοὺς ἔσεσθαι τῶν ἡμετέρων φύλακας διὰ τὴν πρὸς τὸν θεὸν εὐσέβειαν, καὶ μαρτυρουμένους δ᾽ αὐτοὺς ὑπὸ τῶν προγόνων εἰς πίστιν οἶδα καὶ προθυμίαν εἰς ἃ παρακαλοῦνται· βούλομαι τοίνυν καίπερ ἐργώδους ὄντος τοῦ μεταγαγεῖν ὑποσχομένους νόμοις αὐτοὺς χρῆσθαι τοῖς ἰδίοις. ὅταν δ᾽ αὐτοὺς ἀγάγῃς εἰς τοὺς προειρημένους τόπους, εἴς τ᾽ οἰκοδομίας οἰκιῶν αὐτοῖς δώσεις τόπον ἑκάστῳ καὶ χώραν εἰς γεωργίαν καὶ φυτείαν ἀμπέλων, καὶ ἀτελεῖς τῶν ἐκ τῆς γῆς καρπῶν ἀνήσεις ἐπὶ ἔτη δέκα […]. *Cf. 2 Maccabees* 8:20.

82 *1 Maccabees* 10:36–37: καὶ προγραφήτωσαν τῶν Ιουδαίων εἰς τὰς δυνάμεις τοῦ βασιλέως εἰς τριάκοντα χιλιάδας ἀνδρῶν, καὶ δοθήσεται αὐτοῖς ξένια, ὡς καθήκει πάσαις ταῖς δυνάμεσιν τοῦ βασιλέως. καὶ κατασταθήσεται ἐξ αὐτῶν ἐν τοῖς ὀχυρώμασιν τοῦ βασιλέως τοῖς μεγάλοις, καὶ ἐκ τούτων κατασταθήσονται ἐπὶ χρειῶν τῆς βασιλείας τῶν οὐσῶν εἰς πίστιν· καὶ οἱ ἐπ᾽ αὐτῶν καὶ οἱ ἄρχοντες ἔστωσαν ἐξ αὐτῶν καὶ πορευέσθωσαν τοῖς νόμοις αὐτῶν, καθὰ καὶ προσέταξεν ὁ βασιλεὺς ἐν γῇ Ιουδα. See Launey 1987, I, 552; Goldstein 1989, 313–314; Shatzman, 1991, 23.

83 For instance, *1 Maccabees* 3:10; 3:15; 6:21 ff. See Juster 1914, II, 268 note 1.

84 Josephus, *Ant. Iud.*, XII, 289, 299, 303, 362–366. Later on, when King Demetrius II found himself in dire straits, having been abandoned by his troops, Jonathan sent three thousand Jews to Antioch to aid him (*1 Maccabees*, 11:44; Josephus, *Ant. Iud.*, XIII, 133–144).

4 Apologetics and Historical Reality

Josephus' portrayal of Jews as good and loyal soldiers in the service of foreign armies that appears throughout his works[85] can be seen in a particularly significant passage of his *Contra Apionem*:

> Alexander's opinion of the Jews of Alexandria was shared by Ptolemy, son of Lagus. He entrusted the fortresses of Egypt to their keeping, confident of their loyalty and bravery as guards; and, when he was anxious to strengthen his hold upon Cyrene and the other cities of Libya, he sent out a party of Jews to settle there.[86]

This reputation as good soldiers, which is not especially prevalent in other sources mentioning Jews at the time, may have its apologetic origins in Josephus' exaltation of his people's ancestral virtues.[87] However, the author's intentions notwithstanding, we cannot ignore the abundance of sources that bear witness to the frequent establishment of Jewish military colonies[88] and the

85 For instance, Josephus, *Contr. Ap.*, II, 49; *Ant. Iud.*, XI, 318 and 339; XII, 147–150. See Launey 1987, I, 544; Feldman 1993, 88–89.

86 Josephus, *Contr. Ap.*, II, 44: ὅμοια δὲ Ἀλεξάνδρῳ καὶ Πτολεμαῖος ὁ Λάγου περὶ τῶν ἐν Ἀλεξανδρείᾳ κατοικούντων ἐφρόνησεν· καὶ γὰρ τὰ κατὰ τὴν Αἴγυπτον αὐτοῖς ἐνεχείρισε φρούρια πιστῶς ἅμα καὶ γενναίως φυλάξειν ὑπολαμβάνων, καὶ Κυρήνης ἐγκρατῶς ἄρχειν βουλόμενος καὶ τῶν ἄλλων τῶν ἐν τῇ Λιβύῃ πόλεων εἰς αὐτὰς μέρος Ἰουδαίων ἔπεμψε κατοικῆσον.

87 These virtues which he, as a Jew, also partook of, were portrayed as a sort of guarantee of his own leadership in his later writings. According to Michael Tuval (2013, 285–286), "In the closing paragraphs of *AJ*, after congratulating himself on his achievements, Josephus says that now he would like to provide some information on his lineage and the events of his life. This account, presented in *Vita*, centred as it is on Josephus' pedigree and his career in public and military service, reflects well what the contemporary Romans would have expected from a public and political figure, and Josephus created it in order to present himself as the best human example which the Judaean πολιτέα, described in *AJ*, had ever produced." Cf. Launey 1987, I, 542. On the historical method and the credibility of Josephus, see now Tropper 2016, 84–88.

88 Although certain terms such as κατοικία experienced a certain semantic evolution over time, their persistence in Roman-era epigraphy can be interpreted as a remote reminiscence of the origins of certain Jewish communities in the eastern Diaspora as Hellenistic-period military colonies. For instance, we could point out the inscription of *Aurelia Augusta* from Hierapolis (Phrygia), dated to the mid-2nd century CE (*IJO II*, No. 205 = *CIJ II*, No. 775). Even though its latest editor (Walter Ameling) does not deny that the term κατοικία found in this inscription came from the system of military colonies implemented in the Hellenistic kingdoms (*IJO II*, 433–434), its meaning had drifted towards that of "community" or συναγωγή by Roman times (*IJO II*, 435). This, however, does not rule out its remote military origins in the Hellenistic period, as with other Jewish commu-

military service of Jews in Hellenistic armies,[89] despite the fact that certain religious rules in Judaism may have conflicted with military life.[90] Being aware of this fact, Josephus himself provides a possible explanation when he describes, on several occasions, the terms offered by Hellenistic kings to entice Jews into their service,[91] as they may have deemed them to be good and loyal soldiers.[92] This historian even quotes Nicholas of Damascus to show that the Jews who participated in the campaign led by Antiochus VII against the Parthians enjoyed a number of religious privileges:

> And when Antiochus undertook an expedition against the Parthians, Hyrcanus set out with him. On this we have the testimony of Nicholas of Damascus, who writes as follows. "After defeating Indates, the Parthian general, and setting up a trophy at the Lycus River, Antiochus remained there two days at the request of the Jew Hyrcanus because of a festival of his nation on which it was not customary for the Jews to march out." Nor does he speak falsely in saying this; for the festival of Pentecost had come round, following the Sabbath, and we are not permitted to march either on the Sabbath or on a festival.[93]

---

nities in the eastern Diaspora (see Cappelletti 2006, 40; Roux 2020, 7, 12–15). For example, we know that Ptolemy Lagus sent out a party of Jews to settle in Cyrene (Josephus, *Contr. Ap.*, II, 44) and, according to Strabo, the Jews of that city in the time of Sulla formed one of the four classes of the inhabitants (*apud* Josephus, *Ant. Iud.*, XIV, 115).

89 Barclay 1996, 20ff. and 34ff.

90 For instance, Josephus, *Ant. Iud.*, XIV, 10, 12. See Lee 2007, 229 note 1; Grabbe 2008, 221. Josephus (*Contr. Ap.*, 190–191; II, 219, 228, 234) insists on his people's great reverence towards its religious rules to the point of preferring to choose death "rather than repudiate the faith of their forefathers" (μὴ ἀρνούμενοι τὰ πάτρια). Nevertheless, it does not seem that Jews enlisted in Hellenistic armies were able to observe the Sabbath during military marches (Josephus, *Ant. Iud.*, XIII, 252). See Nodet 1997, 89. On the difficulties caused by the Sabbath year (*shmita*) for military life, *1 Maccabees* 6:50–54. See Bar-Kochva 1979, 265 note 3 and 295 note 14. Marcel Launey (1987, I, 551) merely states that "nous ignorons comment les soldats juifs des Ptolémées conciliaient, en campagne et dans les garnisons, les exigences du service avec les règles de leur religion, et spécialement l'observation du Sabbat." Max Radin (1915, 179–181) found it hard to believe that the Ptolemies would have employed as many Jews as they did as mercenaries if they would not fight on the Sabbath. Louis H. Feldman thinks that "the Jewish mercenaries in the army of the Ptolemies may have been unobservant, as many Egyptian Jews were in other respects" (1993, 508, note 88). Cf. Feldman 1960.

91 It is difficult not to see a literary *topos* in this insistent repetition (Batsch 2005, 252–253).

92 For instance, *Ant. Iud.*, XII, 150–152. Marcel Launey concludes that neither in the Ptolemaic forces nor in the Seleucid armies was the ruler cult obligatory (1987, II, 875–1000, especially 923).

93 Josephus, *Ant. Iud.*, XIII, 250–252: [...] καὶ ποιουμένῳ τὴν ἐπὶ Πάρθους αὐτῷ στρατείαν συν-

Based on this text, Christophe Batsch deduced that Jewish military service in Hellenistic armies did not prevent them from observing the Sabbath rest.[94] We must not forget, however, that these Jewish troops were allied auxiliaries (ἐπί-κουροι) and that they fought for King Antiochus under the terms of the "friendship and alliance" forced on John Hyrcanus.[95] In view of these circumstances, it is possible that Nicholas of Damascus noted this exceptional respect towards Jewish religious rules precisely on account of their exceptionality. There are, in fact, insufficient grounds to assume that Jewish service in Hellenistic armies was always subject to respect towards the religious rules of Judaism,[96] and we likewise cannot assume that all Jewish soldiers strictly observed them without renouncing their Jewishness. Starting from his own religious reality, it is quite possible that Josephus was forced to justify behaviour that did not correspond to the uniform notion of Judaism that was beginning to take hold in certain rabbinic circles of his age, refusing to acknowledge the diversity of religious currents and practices that existed and would continue to exist within Judaism.[97] On this subject, it is worth clarifying the notion of Jewish identity with regards to the observance of certain religious rules, taking their historical context into account. According to Haggai Olshanetsky, whom I fully agree with, "there were also different and divergent interpretations of *Kashrut*, and debates about how these laws should be adhered and preserved. The variety of interpretations, behaviours, prayers, artefacts, and so on, imposes the only probable definition for Jews in the Hellenistic era: a Jew is one born to Jewish parents, or one who declares himself to be Jewish. All should be defined, characterized, recognized and accepted as Jews. These include the *Hassidim*, who adhered to a strict conservative abidance of fundamental Torah laws and pro-

---

εξώρμησεν Ὑρκανός. μάρτυς δὲ τούτων ἡμῖν ἐστιν καὶ Νικόλαος ὁ Δαμασκηνὸς οὕτως ἱστορῶν· "τρόπαιον δὲ στήσας Ἀντίοχος ἐπὶ τῷ Λύκῳ ποταμῷ νικήσας Ἰνδάτην τὸν Πάρθων στρατηγὸν αὐτόθι ἔμεινεν ἡμέρας δύο δεηθέντος Ὑρκανοῦ τοῦ Ἰουδαίου διά τινα ἑορτὴν πάτριον, ἐν ᾗ τοῖς Ἰουδαίοις οὐκ ἦν νόμιμον ἐξοδεύειν." καὶ ταῦτα μὲν οὐ ψεύδεται λέγων· ἐνέστη γὰρ ἡ πεντηκοστὴ ἑορτὴ μετὰ τὸ σάββατον, οὐκ ἔξεστι δ' ἡμῖν οὔτε τοῖς σαββάτοις οὔτ' ἐν τῇ ἑορτῇ ὁδεύειν.

94 Batsch 2005, p. 254.

95 See Bar-Kochva 1979, 48.

96 We could assume that there exceptions were made for ethnically homogeneous units. According to A. Kasher (1978, 59–60), "logically, the right to adhere to ancestral laws while serving in the army was common to soldiers in consolidated units based on ethnical and religious characteristics." Cf. Batsch 2005, 253.

97 See Talmon 1991. After having spent more than twenty years in the Diaspora, Josephus himself felt drawn towards the positions of the rabbinic movement which supposedly was becoming prominent in the 90s of the first century, gradually accepting their ideals of Torah-study, respect for ancestral tradition, and meticulous observance of the Law. See Tuval 2013, 257–259.

vided the manpower to the rebels against Antiochus, as well as the *Mityavnim*, who embraced Hellenism and were ready to abandon old Jewish strictures and favour state laws to Torah laws and also sought to popularize Hellenism among their brethren."[98]

98 Olshanetsky 2015, 21. Cf. Barclay 2002. See also the interesting chapter six of Steven Weitzman 2017, 207–244: "Hellenism and Hybridity. Did the Jews Learn How to be Jewish from The Greeks?"

CHAPTER 2

Jewish Exemptions from Military Service in the Late Republic and the Augustan Principate

1 Jews and the Recruitment of *Auxilia*

The Roman Republic never had a standing army. Soldiers, be they citizen legionaries, Latin and Italian *socii*, or *auxilia externa*, were levied at need, usually annually. Especially during the last two centuries of the Republican period, the Roman State made use of troops from outside of Italy, i.e. from peoples who were not part of the Italian *socii ac nomen Latinum*.[1] These soldiers served in certain units known as *auxilia externa*, although the term is used with little regularity. They are usually described by the sources in diverse ways: typically by ethnic and/or type of soldiers, e.g. *funditores*.

These *auxilia externa* (ἐπίκουροι or σύμμαχοι), who sometimes appear in literary sources,[2] seem to be a somewhat exceptional precursor to the future Roman auxiliary troops, the alae and cohorts.[3] The notion of foreign nations collaborating as allies of the Roman army seems to have been accepted and commonplace by the time Varro and Livy wrote their works in the closing decades of the 1st century BCE.[4] The Augustan reforms simply confirmed this practice and granted it legal support. The professionalization of the army, including that of the *externae gentes*, would put an end to the dangerous habit of resorting to mercenaries.[5] Furthermore, this allowed Romans to make sure that a *civis romanus* would be at the head of the new regular troops as an officer.[6] Conversely, we are unaware of the military chiefs or leaders of *externae gentes* from the Republican period. In Imperial times it could happen that, under specific circumstances, Romans may have seen the need to include units of

1 For instance, Polybius, VI, 31, 9; Cicero, *2 Verr.*, V, 60; *Flac.*, 27; Plutarch, *Crass.*, XVII, 9. See Prag 2011, 17; Gauthier 2020.

2 Festus, 16L; Varro, *Ling.*, V, 90; Livy, XXII, 37, 7–8. See Prag 2011, 16–17.

3 Cheesman 1914, 7–25; Saddington 1982, 49; Goldsworthy 1996, 22.

4 For instance, Livy, *Per.*, 72, 6: [...] *auxilia deinde Latini nominis et exterarum gentium missa populo Romano*. See Kendall 2013, 740.

5 Saddington 1982, 86. Recruited auxiliaries served under the same conditions and were fed the same diet as the legionary army: see Roth 1999, 16–17, 84–85.

6 Keppie 1984, 48–51.

 DOI:10.1163/9789004507258_004

soldiers from foreign nations who specialized in specific forms of combat or in the use of weapons that were uncommon in the Roman army.[7] These troops were known as *numeri*,[8] a name which points to their irregular nature and which may be compared to the Republican *auxilia externa*, at least in principle. Indeed, like the *auxilia externa*, they were made up from soldiers of a single ethnicity[9] and they maintained their people's characteristic form of combat.[10] Unlike the cavalry *alae* (ἴλαι) and infantry *cohortes* (σπεῖραι), they did not have a specific amount of personnel.[11]

During the late Republican period, there were times when tumultuary levies were carried out in the provinces when faced with a military emergency or to carry out specific military campaigns authorized by a *senatus consultum*.[12] From the 2nd century BCE onwards, it was very common to carry out minor recruitment drives among the *exterae nationes* to wage local military campaigns outside Italy.[13] Despite the fact that such troops are almost never attested as participating in a triumph at Rome and that their recruitment (*dilectus*) gave rise to much criticism among Roman intellectual circles at the time,[14] the fact is that everything points to the fact that the use of the *auxilia externa* (or *extera*) became increasingly widespread.[15] In the days Julius Caesar and, especially, from the time of Augustus onwards, auxiliaries (*peregrini*) could obtain Roman citizenship after a minimum of twenty-five years of military ser-

7 See Cheesman 1914, 65–66; Saddington 1982, 25, 49 and *passim*.

8 Southern 1989.

9 For example, *Aetoli, Ascalonitae, Canatheni, Ituraei, Damasceni, Sebasteni, Tyrii*, etc. See Schürer 1973, I, 362; 1979, II, 95; Cheesman 1914, 8–9 and 85–90; Kraft 1951, 60–61; Speidel 1975; Spaul 2000, 9 and 437; Myers 2010, 177; Broughton 2019, 4.

10 Phang 2012, 44.

11 In fact, the term *numerus* has the generic meaning of "contingent of troops." See Ureche 2017, 56.

12 Appian, *Iber.*, 38, 156; *Mith.*, 11, 17; Livy, XXXV, 23, 3–9; XXXVII, 2, 8; XXXVII, 57, 5–6; Diodorus Siculus, *B.H.*, 24/25, 2, 18; 26, 3–4; Velleius Paterculus, II, 42, 1–3; Caesar, *Bell. Gall.*, I, 7; III, 20; V, 1. See Prag 2011, 17–18.

13 Prag 2011, 20; Phang 2012, 169; Gauthier 2020, 290–295.

14 Livy, XLIV, 20, 6; Appian, *Iber.*, 48, 201–203; *Mithr.*, 19, 74–75; Frontinus, *Strat.*, II, 7, 3; II, 7, 8; Memnon, *FgrHist.*, 434, F22, 6–8; Cicero, *Fam.*, 104 (XV, 1), 5; *Att.*, V, 18, 2; Caesar, *Bell. Gall.*, III, 20. See Prag 2011, 24.

15 According to Faber (2013, 116), during the early Principate, this military service was opened up to non-Romans on a grand scale. In fact, from the sixties of the first century CE onwards, auxiliary forces "are no longer (if indeed they ever really were) helping forces ancillary to the legions; rather, they are a significant source of power in their own right" (Haynes 2013, 59). Cf. Phang 2012, 79.

vice (*honesta missio*),[16] at which time, starting from the mid-1st century CE, they were granted an official diploma as proof thereof.[17]

The exemptions from military service that were sporadically granted by some Roman generals to Jews during the Civil Wars which would ultimately topple the Republic implies the real possibility, if not the compulsory nature of recruitment prior to these exemptions.[18] In such circumstances Jews, like all other contingents from *exterae gentes*, would have been a part of the auxiliary troops.[19] Nevertheless, we are unaware of when Rome began to raise troops from allied peoples in the eastern Mediterranean.

Even though the debate on the terms and historical extent of the alleged treaties signed by the Hasmonean rulers with Rome from 161 BCE onwards remains open, it seems that the Roman Senate acknowledged the Maccabean State when it had established its independence by its own effort as a "friend and ally".[20] The recognition of Hasmonean Judaea as *socius et amicus populi Romani* through an accord that clearly can only be categorized as *foedus aequum*, implied the obligation of both parties to provide each other military assistance in the event of a defensive or offensive war.[21] Nevertheless, the sources do not record any Roman military interventions in support of the Hasmoneans against the Seleucid kingdom.[22] It seems that Rome did not get involved in any military conflict in the Hellenistic East until the annexation of Asia as

16 Phang 2012, 4.

17 On military diplomas, see Speidel and Lieb (eds.) 2007 and Perea Yébenes 2009.

18 If, as Josephus states, (*Ant. Iud.*, XIV, 204), Caesar decreed that "no one, whether magistrate or pro-magistrate, praetor or legate, shall raise auxiliary troops in the territories of the Jews," (ὅπως μηδεὶς μήτε ἄρχων μήτε ἀντάρχων μήτε στρατηγὸς ἢ πρεσβευτὴς ἐν τοῖς ὅροις τῶν Ἰουδαίων ἀνιστὰς συμμαχίαν), it seems obvious that, until then, the Jews were subject to compulsory recruitment. Cf. Nodet, 1997, 89.

19 Cicero, *Fam.*, 104 (XV, 4), 3: [...] *populorum liberorum regumque sociorum auxilia voluntaria* [...]. See Applebaum 1971, 181; Gruen 2002, 87; Olshanetsky 2018, 11.

20 Cimma 1976, 291 ff.; Baronowski 1982; Sherwin-White 1984, 49, 70 ff.; Goldstein 1989, 310–311; Gabba 1999, 115; Shatzman 1999, 49; Seeman 2013, 115 ff.; Zollschan 2017, 67 ff. and 82; Canfora 2021, 10–14. According to Zvi Yavetz (1997, 89), "[...] Die Römer zogen es vor, im Nahen Osten kleine Klientelstaaten wie Pergamon, Bithynien und Judäa zu fördern. Rom war es lieber, mit einem geschwächten, aber selbständigen Ägypten zu verhandeln als mit einem Riesenreich wie dem der Seleukiden zu tun zu haben [...]."

21 *1 Maccabees* 8:17–32; *2 Maccabees* 4:11; 11:34–38; Josephus, *Ant. Iud.*, XII, 415–419; XIII, 259–266; *Bell. Iud.*, I, 38; Diod. Sic., 40, 2; Justin, *Philippics*, 36, 3, 9. See Juster 1914, II, 269, note 2; Gruen 1984, I, 45–46, 54–96; II, 748–753; Sherwin-White 1984, 71; Baltrusch 2002, 92–98; Wilker 2008, 197; Shatzman 2012; Seeman 2013, 113–118; Rocca 2014a, 267–270 (= Rocca 2014b, 54–57); Zollschan 2017; Wilker 2018, 129.

22 Even though the alliance with Rome did not provide Judaea with any immediate benefits and there was still a long and difficult road ahead before they achieved effective political

a new province of its empire in 133 BCE and the successive rebellion of Aristonicus, which was immediately crushed in 129 BCE.[23] In fact, until the end of the Mithridatic Wars (88–63/62 BCE) and the creation of the province of Syria (63 BCE), the Romans did not feel the need to preserve peace and order in the East.[24] It was then that the Roman Senate granted provincial governors the right of levying in case of need Roman citizens, provincials within their provinces, and allies to bolster the legionary troops assigned to them.[25] After the conquest of Jerusalem by Pompey in 63 BCE[26] and the appointment of Hyrcanus II as high priest and autonomous ethnarch of Judaea, Galilee, Peraea and some districts of Idumaea (albeit under the supervision of the governor of Syria),[27] Jews were, in theory, at the Roman authorities' disposal to be recruited if required.[28] In 57 BCE, Aulus Gabinius, the governor of Syria, gathered forces for an invasion of Parthia, which was interrupted by insurrections of the Jews led by Alexander, the fugitive son of king Aristobulus, who had been deposed by Pompey in favour of Hyrcanus II, and by his mission to restore Ptolemy of Egypt to his throne.[29] Once M. Licinius Crassus took over the province in 55 BCE, he invaded Parthia with a force of seven legions, 4,000 horses, and about 4,000 light-armed men, probably allied troops.[30] After the battle of Pharsalia (48 BCE), the province of Syria pledged its unwavering support to Caesar. Josephus tells that Antipater of Idumaea, acting under orders from Hyrcanus II, provided the Romans with valuable information and "also won over the Jews above Pelusium to his side and made them his allies to act as guards of the entrances to Egypt."[31] According to the Jewish historian, who quoted Strabo

independence, formal recognition of the kingdom by the greatest power in the world was decisive against Seleucid power. See Yavetz 1997, 92; Zollschan 2017, 63 ff.

23 Rocca 2014a, 273 (= Rocca 2014b, 56).

24 Rocca 2014a, 284 (cf. Rocca 2014b, 68). According to Thomas Broughton (2019, 7), "Rome was now for the first time in direct contact with Parthians, who controlled Mesopotamia and continually interfered in the affairs of the various client-kingdoms of Armenia and of Syria."

25 Cicero, *Att.* 111 (V, 18); *Fam.* 104 (XV, 1).

26 Bellemore 1999; Troiani 2012; Cameron 2018; Canfora 2021, 23–27, 31–36, and *passim*.

27 Josephus, *Bell. Iud.*, I, 153; *Ant. Iud.*, XIV, 73; XX, 244. See Burr 1972; Cimma 1976, 217; Smallwood 1976, 27; Shatzman 1991, 129–131; Gabba 1999, 93–96; Troiani 2012, 94–95; Seeman 2013, 253.

28 Schürer, 1973, I, 317; Zeichmann 2018, 4.

29 Josephus, *Ant. Iud.*, XIV, 83. See Gabba 1999, 98.

30 Plutarch, *Crass.*, 20. See Gabba 1999, 99; Seeman 2013, 289 ff.; Broughton 2019, 7.

31 Josephus, *Ant. Iud.*, XIV, 99: […] Γαβινίῳ μέντοι κατὰ τὴν στρατείαν ἣν ἐφ' Ὑρκανὸν ἐστείλατο Ἀντίπατρος ὑπηρέτησεν σῖτον καὶ ὅπλα καὶ χρήματα, καὶ τοὺς ὑπὲρ Πηλούσιον τῶν Ἰουδαίων

to support his information with other authorized sources (as he often did), an army of Jews, Arabs, and Syrians under Mithridates of Pergamum and Antipater of Idumaea, then governor of Judaea, came to relieve Caesar in Alexandria.[32] According to Strabo,[33] it seems that, on that occasion, the contingent from Judaea was 3,000 men strong, and they would later be joined by Jews settled in the Delta area who had been recruited by Antipater himself.[34] Hyrcanus II would subsequently send another army commanded by his sons to assist Caesar's supporters against pro-Pompeian Caecilius Bassus.[35]

Cassius Longinus and Marcus Antony in succession levied auxiliary troops from Syria and Asia Minor as contingents of the allied kings.[36] We know that, according to Plutarch, Antony initially had Jewish auxiliaries sent by Herod in his army when he confronted Octavius.[37] However, as the balance tilted in Octavius' favour, the Jewish king began to collaborate with him.[38] In fact, he immediately sent military assistance to Quintus Didius, governor of Syria, to fight Cyzicus' gladiators, who had rallied to Antony's cause.[39] From that point onwards, military cooperation between the Herodian kingdom and "Rome" would remain constant.[40] It is worth pointing out that, in 30 BCE, Herod supplied the armies Octavius sent to Egypt through Syria[41] and, five years later, he provided Aelius Gallus, prefect of Egypt, with five hundred men as rein-

οὗτος αὐτῷ προσηγάγετο καὶ συμμάχους ἐποίησεν φύλακας ὄντας τῶν εἰς τὴν Αἴγυπτον ἐμβολῶν [...].

32 Schürer 1973, I, 270–271; Grainger 2013, 39.

33 Josephus, *Ant. Iud.*, XIV, 139; cf. *Bell. Iud.*, I, 187. Elsewhere, Josephus states that there were 1,500 soldados (*Ant. Iud.*, XIV, 193) and Nicholas of Damascus states that there were 2,000 in his plea to Agrippa (*Ant. Iud.*, XVI, 52).

34 *Ant. Iud.*, XIV, 127–139; *Bell. Iud.*, 187–192. See Applebaum 1989b, 54; Shatzman 1991, 133, 139, and 283; Maier 1992, 82; Feldman 1993, 93; Barclay 1996, 291; Gabba 1999, 100; Grant 1999, 58–59; Rodrigues 2007, 281; Roth 2007, 417; Canfora 2012.

35 Josephus, *Ant. Iud.*, XIV, 269. Cf. Juster 1914, II, 269, note 2.

36 Josephus, *Bell. Iud.*, I, 202–203, 216–217, 225, 324, 346–346; *Ant. Iud.*, XIV, 271–272, 449, 468–469; Appian, *B. C.*, 5, 87; Cicero, *Att.*, 363 (XIV, 9); Plutarch, *Anton.*, 37, 3; Cassius Dio, 47, 26–27. See Broughton 2019, 8.

37 Plutarch, *Anton.*, 61. According to Josephus (*Ant. Iud.*, XIV, 439 ff. and *Bell. Iud.*, I, 322), Herod also sent cavalry and infantry to support Antony during his siege of Samosata, on the Euphrates. See Jacobson 2001, 26; Saddington 2009, 319; Gauthier 2020, 293.

38 Feldman 1993, 90.

39 Josephus, *Ant. Iud.*, XV, 195; *Bell. Iud.*, I, 392. Cf. Cassius Dio, 51, 7 and Plutarch, *Anton.*, 71–72. See Shatzman 1991, 168; Saddington 2009, 319–320.

40 See Juster 1914, II, 269–271 note 2; Saddington 2009, 311 and *infra*, Jewish Troops in Roman Service (pp. 45–61).

41 Josephus, *Ant. Iud.*, XV, 196–201; *Bell. Iud.*, I, 394–396. See Shatzman 1991, 169; Richardson 1999, 172–173; Günther 2007, 89; Roth 2007, 415.

forcements for Roman troops in their expedition against the Arabs.[42] Part of the Herodian army was eventually incorporated into the Roman *auxilia*.[43] It is very likely that, apart from Ituraean archers,[44] the Jewish king also had other light infantry units composed of Jews that later served as *auxilia* in the Roman army.[45] It is also very likely that Herod's Jewish army, of which the Idumaeans converted to Judaism by John Hyrcanus (ca. 130 BCE) were undoubtedly a part,[46] may have been trained and organized in the Roman fashion.[47] Josephus utilizes various terms in describing Herod's infantry that could imply that it was organized along Roman lines: χιλίαρχος for *tribunus militum*, ταξίαρχος for *centurio*, and σπεῖρα for *cohors*.[48] In this regard, Josephus speaks of five Roman cohorts (σπεῖραι) and five Jewish ones operating together.[49] Moreover, there were also some officers with Roman names, such as the military tribune Volumnius, who commanded Herodian troops.[50]

It is unsurprising that, after the passing of Herod the Great (4 BCE), some of his elite troops, such as the *cohortes* and *ala Sebastenorum*, ended up being definitively included in the Roman army as auxiliary forces[51] that would end

42 Josephus, *Ant. Iud.*, XV, 317; Strabo, XVI, 4, 23. See Shatzman 1991, 183–184; Alston 1995, 29; Rocca 2008, 133, 148, 151; Saddington 2009, 320.

43 Shatzman 1991, 198; Rocca 2008, 59, 133, 146; Zeichmann 2018, 4.

44 Josephus, *Ant. Iud.*, XIV, 452, 468 and *Bell. Iud.*, I, 329, 346. Various Ituraean light units serving as *auxilia* in the Roman Imperial army are known from epigraphical sources. These are the *Cohors I Augusta Ituraeorum S.* that served in Pannonia and Dacia, the *Cohors II Ituraeorum* that served in Germania Superior and Dacia, the *Cohors II Ituraeorum E.* and the *Cohors III Ituraeorum* that served in Egypt, the *Cohortes IV, V, VI Ituraeorum* that are supposed to have existed on behalf of the *Cohors VII Ituraeorum* that served in Egypt too. See Cheesman 1914, 182; Rocca 2008, 146–147, note 61.

45 The Idumaeans settled in Trachonitis served as archers (Josephus *Ant. Iud.*, XVI, 292 and *Bell. Iud.*, II, 58). See Shatzman 1991, 175; Rocca 2008, 147.

46 Josephus, *Ant. Iud.*, XIII, 254–258; XV, 254. It seems that the Idumaeans identified as Jewish (*Bell. Iud.*, IV, 270–282) and, even though they were scorned for their pagan origins, the Jews themselves acknowledged them as such (*Ant. Iud.*, XIV, 403). According to Katell Berthelot (2019, 118), "even in the absence of Judaean citizenship in the Hasmonean state, the integration of the Idumaeans in the Hasmonean society and the careers of some Idumaeans at the Hasmonean court, as illustrated by that of Antipater, Herod's father, can be considered comparable to the integration of the Italian allies by the Romans." Cf. Shatzman 1991, 58.

47 Roth, 2007, 413; *Idem* 2019, 89.

48 Shatzman 1991, 206–208; Saddington 2009, 314–315. In fact, "these terms are likewise utilized by Polybius, by Asclepiodotus, and in Ptolemaic papyri to indicate units of heavy infantry that were organized and equipped like the Roman legions" (Rocca 2008, 147).

49 Josephus, *Bell. Iud.*, I, 301. See Gabba 1999, 136.

50 Josephus, *Bell. Iud.*, I, 535; *Ant. Iud.*, XVI 332, 356–366. See Gracey 1986, 314; Rocca 2019, 86.

51 Josephus, *Ant. Iud.*, XX, 122; *Bell. Iud.*, II, 236; *Acts*, 27:1. See Shatzman 1991, 185, 194; Millar

up playing a very active role in the Jewish war.[52] According to Josephus, these 3,000 Samaritans were a part of Herod's army, organized into five cohorts of infantry (σπεῖρα Σεβαστή) and a wing of cavalry (ἴλη Σεβαστηνῶν).[53] The name *Sebaste*, honouring Augustus, shows it did not predate January 27 BCE.[54] One of these cohorts and the cavalry *ala* are attested in inscriptions.[55] These troops were commanded by Rufus and Gratus, both names of Latin origin.[56]

2 Military Exemption as a Jewish Privilege

The earliest information on the possible military service of Jews in the Roman army comes from historian Josephus, who precisely refers to the exemptions from service granted by the Roman authorities at a time when the State, wracked by civil conflicts, was in critical condition, and the Republic drew its last breath after a protracted agony.[57] It is certainly significant that such apparently counter-intuitive decisions would have been adopted in such a context of armed conflict and political instability. It seems that the diplomatic game and shifting positions according to political convenience may have exerted an influence on events and, therefore, on any measures that were incidentally adopted.[58] The fact remains that, thanks to the pro-Pompeian consul Lucius Lentulus Crus, Jews from the province of Asia who had Roman citizenship were first exempted from service in 49 BCE.[59] The decree, which was specifically addressed to the Jews of Ephesus, read as follows, according to Flavius Josephus:

1993, 60; Saddington 2009, 311, 313; Haynes 2013, 117; Broughton 2019, 18 and 23; Roth, 2019, 88–89, 147. According to Samuel Rocca (2008, 147 note 64), "there was an *Ala Sebastenorum* that served in Mauretania Caesarensis, and a *Cohors I Sebastenorum M(iliaria)* that continued to serve in the Land of Israel. The fact that *Cohors I* is mentioned means that *Cohors II* also existed, thus 3,000 soldiers, the same number mentioned by Josephus."

52 Josephus, *Bell. Iud.*, III, 65–66; *Ant. Iud.*, XX, 122. See Millar 1993, 60; Zeichmann 2018, 4; Broughton 2019, 23.

53 *Ant. Iud.*, XVII, 266. 275–276. 283. 294; XX, 122; *Bell. Iud.*, II, 52. 58. 63. 74; III, 65–66.

54 Cf. *Acts* 27:1. See Shatzman 1991, 180, note 36.

55 *DMIPERP* § 153, § 154, § 213, § 219, §§ 221–245. See Spaul 2000, 29, 438, 453, 504, 521–522.

56 Josephus, *Bell. Iud.*, II, 52; *Ant. Iud.*, XVII, 266. See Gracey 1986, 314; Rocca 2008, 147; Saddington 2009, 313–314.

57 Ritter 2015, 199.

58 González Salinero 2003, 45.

59 Schürer 1986, 22; Juster 1914, 275; Kasher 1985, 77; Sartre 1991, 394; Ritter 2015, 204 note 14, 212, and 217–218.

> Those Jews who are Roman citizens and observe Jewish rites and practise them in Ephesus, I released from military service before the tribunal on the twelfth day before the Kalends of October [September 19] in consideration of their religious scruples, in the consulship of Lucius Lentulus and Gaius Marcellus […].[60]

It seems that the application of this decree was not restricted to Jews from the capital, but rather also applied to the entire province. This is shown by the words of Josephus himself when he states that, when he received the order, the legate and propretor Titus Ampius Balbus assumed that L. Lentulus' decision affected all Jews in Asia,[61] and it was likewise extended to those who lived in Delos and in Sardis, this time under the responsibility of legate Marcus Piso and proquaestor and propretor Lucius Antonius.[62]

Even though the Marian reforms had put an end to volunteer service as the most common method of recruitment for the legions,[63] Asian Jews who held Roman citizenship were able to avoid the draft by virtue of the decree passed by L. Lentulus.[64] We must assume that those Jews who had not obtained Roman citizenship would have been integrated into the auxiliary contingents which, as we pointed out earlier, were drawn from allied peoples and were sometimes as numerous as the legionaries, who were entirely recruited from Roman citizens.[65]

According to Flavius Josephus, religious reasons (δεισιδαιμονίας ἕνεκα) were behind the establishment of a "special situation" for Jewish Roman citizens.[66] The Jewish historian specifically referred to the underlying religious consider-

---

60 Josephus, *Ant. Iud.*, XIV, 228: πολίτας Ῥωμαίων Ἰουδαίους ἱερὰ Ἰουδαϊκὰ ἔχοντας καὶ ποιοῦντας ἐν Ἐφέσῳ πρὸ τοῦ βήματος δεισιδαιμονίας ἕνεκα στρατείας ἀπέλυσα πρὸ δώδεκα καλανδῶν Ὀκτωβρίων Λευκίῳ Λέντλῳ Γαΐῳ Μαρκέλλῳ ὑπάτοις […]. The text appears again, with slight variants, in *Ant. Iud.*, XIV, 234 and 240.

61 Josephus, *Ant. Iud.*, XIV, 230: […] Ἰουδαίους τοὺς ἐν τῇ Ἀσίᾳ Λεύκιος Λέντλος ὁ ὕπατος ἐμοῦ ἐντυχόντος ἀπέλυσεν τῆς στρατείας […]. Cf. Rodrigues 2007, 319; Roux 2020, 23–25.

62 Josephus, *Ant. Iud.*, XIV, 231–232 and 235. Cf. Trebilco 1991, 16–17, 172; Goodnick Westenholz 1994, 68; Varon 2001, 271–272.

63 Smith 1958; Harmand 1967; Gabba 1976; Nicolet 1980, 92–93; Alston 2002, 30–34.

64 As Edith Mary Smallwood (1976, 127–128) has pointed out, the number of Jews affected by the measures must have been very small, and the exemption may have been almost unnoticeable. Nevertheless, L. Lentulus' decision was significant with regards to the emergence of a principle of "tolerance" which, even if it was only formal, constituted a gesture of sympathy towards Judaism. Cf. Juster 1914, I, 142 ff. and II, 274–275; Applebaum 1971, 181; Rajak 1985, 23–25; Grant 1999, 58.

65 Saulnier 1981, 168.

66 Cf. Trebilco 1991, 16–18; van der Horst 2013, 336 (= 2014, 159).

ations in the decree when describing the contents of the exemption granted anew to the Jews by P. Cornelius Dolabella shortly afterwards, in 43 BCE. It is quite possible that, once L. Lentulus left Asia, his measures became null and void, and Jews were once again included in the conscription assigned to that province.[67] At the petition of the Hasmonean High Priest (and ethnarch, a title granted by Caesar)[68] Hyrcanus II,[69] Dolabella, the Caesarean governor of Syria, wrested Asia from the grip of legate C. Trebonius and extended the former military exemption to Jews without Roman citizenship, thus guaranteeing observance of the ritual rules required by their religion:

> [...] Alexander, son of Theodorus, the envoy of Hyrcanus, son of Alexander, the high priest and ethnarch of the Jews, has explained to me that his co-religionists cannot undertake military service because they may not bear arms or march on the days of the Sabbath; nor can they obtain the native foods to which they are accustomed. I, therefore, like the governors before me, grant them exemption from military service and allow them to follow their native customs and to come together for sacred and holy rites in accordance with their law, and to make offerings for their sacrifices; and it is my wish that you write these instructions to the various cities.[70]

We must not forget that, in thanks for the assistance provided by Hyrcanus II and Antipater in a context of dire straits in Egypt during the so-called *bellum alexandrinum*,[71] Julius Caesar himself had acknowledged a number of "privileges" for the Jews in 47 BCE, which included an exemption from the recruitment of auxiliary troops in Judaea:

---

67 Smallwood 1976, 128.

68 The letter sent by Caesar to the Sidonians in 47 BCE (Josephus, *Ant. Iud.*, XIV, 190–195) contains the appointment of Hyrcanus as hereditary ἐθνάρχης and ἀρχιερεύς of the Jews. Cf. Schürer 1973, 272.

69 Josephus, *Ant. Iud.*, XIV, 223–224.

70 Josephus, *Ant. Iud.*, XIV, 226–227: [...] Ἀλέξανδρος Θεοδώρου πρεσβευτὴς Ὑρκανοῦ τοῦ Ἀλεξάνδρου υἱοῦ ἀρχιερέως καὶ ἐθνάρχου τῶν Ἰουδαίων ἐνεφάνισέν μοι περὶ τοῦ μὴ δύνασθαι στρατεύεσθαι τοὺς πολίτας αὐτοῦ διὰ τὸ μήτε ὅπλα βαστάζειν δύνασθαι μήτε ὁδοιπορεῖν ἐν ταῖς ἡμέραις τῶν σαββάτων, μήτε τροφῶν τῶν πατρίων καὶ συνήθων κατὰ τούτους εὐπορεῖν. ἐγώ τε οὖν αὐτοῖς, καθὼς καὶ οἱ πρὸ ἐμοῦ ἡγεμόνες, δίδωμι τὴν ἀστρατείαν καὶ συγχωρῶ χρῆσθαι τοῖς πατρίοις ἐθισμοῖς ἱερῶν ἕνεκα καὶ ἁγίοις συναγομένοις, καθὼς αὐτοῖς νόμιμον, καὶ τῶν πρὸς τὰς θυσίας ἀφαιρεμάτων, ὑμᾶς τε βούλομαι ταῦτα γράψαι κατὰ πόλεις. See Juster 1914, I, 358; Goldenberg 1979, 416; Rabello 1980, 706–707; Schürer 1986, 22, 121; Rodrigues 2007, 319–320, 677 note 236; Krause 2017, 82–83; Olshanetsky 2018, 13.

71 Josephus, *Ant. Iud.*, XIV, 123–139. Cf. Tcherikover 1982, 284; Rocca 2010, 18.

> […] And that no one, whether magistrate or pro-magistrate, praetor or legate, shall raise auxiliary troops in the territories of the Jews, nor shall soldiers be allowed to exact money from them, whether for winter quarters or on any other pretext, but they shall be free from all molestation […].[72]

The Senate likewise ratified Caesar's policy towards the Jews through a *senatus consultum*.[73] For this reason, it is quite likely that they were not exempted from military service (ἀστρατεία) due to Roman awareness of their religion (as Josephus implies time and again), but rather due to their loyalty towards Caesar's cause.[74] In any event, in the light of these precedents, it is unsurprising that the Jews attempted to perpetuate a situation that was apparently very beneficial for them.[75] For this reason, the Jews from several cities in Ionia, who were immersed in a bitter dispute with their Greek fellow citizens due to the latters' lack of respect towards the religious obligations of Mosaic law, turned to the authority of Marcus Vipsanius Agrippa, the famous general and Augustus' son-in-law, in 14 BCE, as he returned with his troops through Asia Minor after his military campaigns in the Bosphorus alongside Herod the Great.[76] The presence of the Jewish king and his known friendship with Agrippa[77] may have led the Jews to expect him to rule in their favour, and they appointed Nicholas of Damascus, a Greek counsellor and historian in Herod's court, as their represen-

---

72 Josephus, *Ant. Iud.*, XIV, 204: […] καὶ ὅπως μηδεὶς μήτε ἄρχων μήτε ἀντάρχων μήτε στρατηγὸς ἢ πρεσβευτὴς ἐν τοῖς ὅροις τῶν Ἰουδαίων ἀνιστὰς συμμαχίαν καὶ στρατιώτας ἐξῇ τούτῳ χρήματα εἰσπράττεσθαι ἢ εἰς παραχειμασίαν ἢ ἄλλῳ τινὶ ὀνόματι, ἀλλ' εἶναι πανταχόθεν ἀνεπηρεάστους. […]. See Mendelssohn 1875, 195–197; Juster 1914, I, 138–149; Askowith 1915, 202; Schürer 1973, I, 272; Saulnier 1981, 178; Rajak 1985, 24; Gaudemet 1988, 241; Maier 1992, 82; Pucci Ben Zeev 1995, 28 ff.; Rodrigues 2007, 283–288; Nodet 2019, 64. As John M.G. Barclay (1996, 291) has stated, it is possible that the acknowledgement of Jews as "friends and allies" may have also benefitted Jewish communities outside Judaea. According to a study carried out by Miriam Pucci Ben Zeev (1995, 29–30), there can be no doubt as to the veracity of the information provided by Flavius Josephus with regards to Caesar's attitude towards the Jews.

73 Josephus, *Ant. Iud.*, XIV, 219–222. As Miriam Pucci Ben Zeev (1994) has demonstrated, the contents and technical details of Josephus' textual quotation of the *senatus consultum* that ratified the privileges granted by Caesar to the Jews prove that it is a copy of the original decree beyond a shadow of a doubt. Cf. Pucci Ben Zeev 2002; Zack 2018, 165–179.

74 Roth 2007, 417.

75 Cf. van der Horst 2014, 146.

76 Josephus, *Ant. Iud.*, XVI, 27 ff. See Juster 1914, II, 275; Smallwood 1976, 140–141; Grant 1999, 76; Barclay 1996, 277–278; Rocca 2019, 227. It seems that the Jews' prerogatives often gave rise to mistrust among the Greeks (Feldman 1993, 94).

77 On Flavius Josephus' idyllic description of the friendly relations between Herod and Agrippa, see Roddaz 1984, 450–463.

tative and spokesman to voice their complaints before the Roman dignitaries.[78] Among the many "vexations" to their legally-acknowledged religious freedom they had had to endure, the Jews stated that they were also "forced to participate in military service" (στρατειῶν ἀναγκαζόμενοι).[79] The claim and the ensuing decision hinged on a single argument: the Jews had enjoyed and should continue to enjoy a privileged position by virtue of their ancestral religious customs.[80] Flavius Josephus describes Agrippa's decision as follows:

> [...] And since they asked that the rights which they had formerly received should not be annulled, he would confirm their right to continue to observe their own customs without suffering mistreatment [...].[81]

3 A Legal Precedent?

According to some authors, Caesar's granting of a "Magna Carta" or "privileged legal situation" that became perpetuated over time due to its subsequent confirmation and inclusion in the Roman legal tradition allowed Jews in general (including those in the Diaspora) to enjoy a special legal status that exempted them from conscription.[82] Shimon Applebaum and Tessa Rajak, among others, rejected this interpretation, suggesting that the "privileges" granted by Caesar had a limited scope and a restrictive value.[83] This can be seen in the fact that, after his death, his measures were ignored, and Jewish religious freedom was openly denied in many parts of the Diaspora. Roman interference was limited to ruling on a number of complaints from certain Jewish communities. In fact, the need to constantly renew the acknowledgement of "Jewish rights" proves, according to this author, that the privileges Caesar granted their religion barely

---

78 Rocca 2019, 227.

79 Josephus, *Ant. Iud.*, XVI, 27–28; cf. *Ant. Iud.*, XIV, 45.

80 Josephus, *Ant. Iud.*, XVI, 35, 41; cf. XII, 126. See Juster 1914, II, 278; Richardson 1999, 270–272; Rajak 1992, 17; Rodrigues 2007, 376. Cf. Rocca 2010, 18.

81 Josephus, *Ant. Iud.*, XVI, 60: [...] ἐπεὶ δὲ ἃ καὶ πρότερον εἰλήφασιν ἄκυρα μὴ γενέσθαι, βεβαιοῦν αὐτοῖς ἀνεπηρεάστοις ἐν τοῖς οἰκείοις διατελεῖν ἔθεσιν [...]. See Rodrigues 2007, 376; Ritter 2015, 220.

82 See Juster 1914, I, 217; McCasland 1943, 65; Wilde 1949, 30; Smallwood 1976, 124, 128, 137, 539, 559, *passim*; Goldenberg 1979, 416; Rabello 1980, 692 and 695; Otzen 1990, 56; Trebilco 1991, 197–198 note 58 and 18; Goodnick Westenholz 1994, 29 and 60; Sartre 2005, 104.

83 In his critical review of the works of Jean Juster, Théodore Reinach (1920, 25) already denounced the excessive importance the author granted to the edicts of Lentulus and Dolabella and their lack of weight as proof that they eventually became generalized.

had any legal value.[84] Moreover, military exemptions were not expressly ratified by Augustus in his general confirmation of Jewish rights in 2 BCE/2 CE.[85] Furthermore, according to Shimon Applebaum, those exemptions were a consequence of the Civil Wars, during which both sides pursued a self-interested policy of conciliation with pro-Roman elements in Asia, and any concessions granted at that time are therefore not evidence of a permanent and universal exemption of Jews from military service in the Roman Empire.[86] Thus, in theory, there were no obstacles to the inclusion of Jews in the *dilectus* in provinces where they were present in large numbers, such as Syria, Asia, Egypt, Cyrenaica or Cyprus.[87]

Nevertheless, the information provided by Josephus and other sources somewhat contradicts the vision put forward by Shim'on Applebaum and Tessa Rajak. It is very likely that Jews were able to preserve their right to live according to their customs in the Roman Empire, both in Judaea and in the Diaspora.[88] At the very least, this is clear from Josephus' description of a verifiable situation in several locations: Alexandria, Ephesus, Miletus, Smyrna, Tralles, Pergamum, Halicarnassus, Laodicea, Sardis, Cyrene, Dora, and the islands of Paros and Cos.[89] The friction that arose in these cities between Jews and local authorities were due to the frequent breaches of the civil status that protected Jews from the "Greek citizen bodies of the host cities."[90] Despite the fact that their "rights and customs" had been, in general, ratified both by Augustus and Claudius,[91] the legal situation of Jews was not always set in stone. As Pieter van der Horst correctly noted, "Roman authorities had granted the Jewish communities certain rights and privileges (e.g., exemption from military service), and the Jews had to struggle for their maintenance, but they apparently had sufficient influence and goodwill to get things done as they wanted."[92] In fact, we do not know

84 See Rajak 1984, 109–115; *Idem*, 1985, 31–32.
85 Josephus, *Ant.*, XVI, 160. See Applebaum 1971, 183 note 2; Saulnier 1981, 182–188; Oppenheimer 2005a, 187; Rocca 2010, 17; *Idem* 2019, 228–229. Tessa Rajak (1992, 17) is mistaken when she states that "the consequent exemption granted by Augustus appeared still in the Theodosian Code (II 8, 6 = VIII, 8, 8; XVI 8, 2)." Constitution II, 8, 6 does not exist, and the other two laws quoted from the Theodosian Code do not refer to the military exemption.
86 Applebaum 1974, 459. Cf. Vian 2001, 273.
87 Applebaum 1971, 181.
88 Pucci Ben Zeev 1995, 31 ff.
89 Josephus, *Ant. Iud.*, XIV, 213–216; 244–246; XVI, 160–165; XIX, 278–285; XIX, 301–312; *CPJ II*, 153 (Alexandria). See McCasland 1943, 67; Grant 1999, 59; van der Horst 2013, 324 (= 2014, 145–146).
90 Smallwood 1999, 177.
91 Josephus, *Ant. Iud.*, XVI, 162–165 y 172; XIX, 280–285 y 290. See Gaudemet 1988, 343.
92 van der Horst 2013, 324 (= 2014, 146).

of any revocations of the religious freedom of Jews, not even at the time of the first revolt against Rome when, according to Josephus, Titus said that the Romans allowed the use of Jewish law: "we maintained the laws of your forefathers and permitted you, not only among yourselves but also in your dealings with others, to live as you willed."[93] The fact that these rights were confirmed on subsequent occasions does not necessarily mean that, as Tessa Rajak argues, they lacked value. Quite the contrary, constant references to this reality could mean that Caesar's concessions were still being considered and that, far from being a mere memory, they had an undeniable legal value.

However, historians who have followed the lead of Jean Juster on the alleged existence of a number of exclusive privileges for Jews in the Roman Empire ignore the fact that other peoples, such as the Greeks, enjoyed similar concessions. As Miriam Pucci ben Zeev pointed out, Romans were inclined, at various times and for different reasons, to acknowledge the right of conquered nations to govern themselves according to their own laws.[94] It seems that, with regards to the Jews, the key factor was the assistance they provided to Caesar during the Civil War.[95] Thus, Jean Juster was right to believe that Caesar's permission to the Jews to live according to their laws had legal value, but he was mistaken in his belief that it was "special". Comparisons with other Greek epigraphic documents (many of which were still unknown at Juster's time) reveal that Roman authorities frequently granted similar permissions to other conquered peoples.[96] We could thus state that Roman policy was characterized by the maintenance of the *status quo ante* with regards to local traditions whenever it was possible (or suitable), and that the case of the Jews was therefore not exceptional. Thus, the guarantees of freedom of worship granted to Jews should not be deemed to strictly be *privilegia* or exclusive concessions.[97] In any event, we do not have enough reliable information to categorically state that, after the Augustan period, these favourable concessions to Jews included exemption from military service or that this exemption remained fully in force

93 Josephus, *Bell. Iud.*, VI, 334: […] ἔπειτα τοὺς πατρίους νόμους ἐτηρήσαμεν, καὶ ζῆν οὐ μόνον καθ' ἑαυτοὺς ἀλλὰ καὶ πρὸς τοὺς ἄλλους ἐπετρέψαμεν ὡς ἐβούλεσθε […].

94 Pucci Ben Zeev 1995, 33.

95 Josephus, *Ant. Iud.*, XIV, 192–195. See Pucci ben Zeev 1995, 35–36. For this reason, at the time of the Roman procurators (from 6 CE onwards, with the exception of the brief reign of Agrippa I, 41–44 CE), there was no compulsory recruitment of Jews for military service in Judaea (see McCasland 1943, 67; Applebaum 1974, 456).

96 The rights the Romans granted to the Greeks also included an exemption from military service (ἀστρατεία). See Pucci ben Zeev 1995, 37; Sherk 1969, 193 note 1; Grant 1999, 58; Millar 1977, 420–434.

97 See Pucci ben Zeev, 1996, esp. 48–49.

throughout the remainder of the Julio-Claudian dynasty.[98] However, even in the unlikely case that it had been so, we must assume that this exemption would have been revoked, at least in what concerns the Jews of Rome and Italy, by Emperor Tiberius in the fifth year of his reign (19 CE).[99]

98 Cappelletti 2006, 61.

99 Juster 1914, II, 276; McCasland 1943, 65; Varon 2001, 273 (though he mistakenly adds Seneca, *Epist.*, 108, 22 as a source on this revocation); Laurenzi 2011, 17.

CHAPTER 3

Jewish Soldiers in the Roman Army during the High Empire

1 Exceptional Recruitment

The repeated claims for the concessions granted to Jews by Roman authorities could lead us to believe that exemption from military service may have remained in force in the eastern part of the Empire, or that, at least, Jewish communities in the eastern Diaspora in Greek-majority cities were able to invoke various legal precedents to avoid local recruitment.[1] Nevertheless, the lack of sources does not allow us to confirm whether or not the alleged prolongation of this hypothetical exemption from military service also applied to Jews in the western Diaspora.[2] In any case, had it existed, it would have been revoked in the year 19 CE by Emperor Tiberius, as shown by his compulsory recruitment of Jews from the City of Rome—or from Italy as a whole, according to Tacitus—,[3] to send them to Sardinia, in an episode described both by Josephus and by Tacitus and Suetonius.[4] Even though Tacitus' text displays a clear anti-Jewish slant,[5] it also provides more details:

> There was also discussion of driving out Egyptian and Jewish rites, and a senatorial decree was passed ordering 4,000 persons of the freedman class who had been infected with such superstition, and who were also

1 Rabello 1980, 743; Sartre 1991, 365. On the question of "local militias", see Le Bohec 2000, 32–33.

2 Panayotov 2004, 61.

3 Note that, when describing the expulsion of astrologers and actors from Rome, Tacitus likewise always talks about expulsions from the whole of Italy (*Ann.*, II, 32; IV, 14; XV, 52; *Hist.*, II, 62). Is this a coincidence?

4 Hudson 1882, 74; Juster 1914, II, 276; Radin 1915, 305–309; Heidel 1920; Smallwood 1956; Abel 1968; Barnett 1974–1975; Smallwood 1976, 202–203; Kasher 1985, 77; Williams 1989; Le Bohec 1990, 22–23; Barclay 1996, 298; Rutgers 1998a, 176–181 (= 1998b, 98–105); Cappelletti 2006, 43; Castelli 2009; Rocca 2010; Shean 2010, 182; van der Lans 2015, 36–37; Wendt 2015, 109–110; Olshanetsky 2018, 13–14.

5 Rokéah 1995; Barclay 1996, 314–315, 362–363; Schäfer 1997, 31–33, 74–75; Gruen 2009 (= 2016). Although René S. Bloch shares the view that Tacitus' portrait is a hostile one, he offers a more nuanced and complex analysis that transcends the simplistic vision of anti-Judaism (2002, 159–176).

 | DOI:10.1163/9789004507258_005

> of appropriate age, to be transported to the island of Sardinia. There they were to be employed in putting down banditry, and if they perished because of the oppressive climate the loss would be slight. The others were under orders to leave Italy unless they cast aside their profane religious observances before a specific date.[6]

At the emperor's behest,[7] the Senate decided, through a *senatus consultum*, to create a military unit by forcibly recruiting 4,000 Jewish freedmen (*libertinum genus*), young and with the proper age (*idonea aetas*, *Iudaeorum iuuentus*) to serve in the army (*per speciem sacramenti*). Even if we were to accept that this decision may have responded to a wish to carry out a social purge,[8] this episode seems to be yet another manifestation of the policy of compulsory and extraordinary conscription to respond to pressing circumstances.[9] Tacitus states that

---

6 Tacitus, *Ann.*, II, 85: *Actum et de sacris Ægyptiis Iudaicisque pellendis factumque patrum consultum, ut quattuor milia libertini generis ea superstitione infecta, quis idonea aetas, in insulam Sardiniam ueherentur, coercendis illic latrociniis et, si ob grauitatem caeli interissent, uile damnum; ceteri cederent Italia, nisi certam ante diem profanos ritus exuissent.* Cf. Suetonius, *Tib.*, 36: "He suppressed foreign cults and the religions of the Egyptians and the Jews, obliging those who practised such rituals to burn their religious garments and all their paraphernalia. The young men of the Jewish people he had sent to regions where the climate was severe, ostensibly on military service. The rest of that people, and others of similar beliefs, he banished from the city, with the penalty of slavery for life if they did not obey [...]" (*Externas caerimonias, Ægyptios Iudaicosque ritus compescuit, coactis qui superstitione ea tenebantur religiosas uestes cum instrumento omni comburere. Iudaeorum iuuentutem per speciem sacramenti in prouincias grauioris caeli distribuit, reliquos gentis eiusdem uel similia sectantes urbe remouit, sub poena perpetuae seruitutis nisi obtemperassent* [...]); Josephus, *Ant. Iud.*, XVIII, 84: "The consuls drafted four thousand of these Jews for military service and sent them to the island of Sardinia [...]" (οἱ δὲ ὕπατοι τετρακισχιλίους ἀνθρώπους ἐξ αὐτῶν στρατολογήσαντες ἔπεμψαν εἰς Σαρδὼ τὴν νῆσον [...]).

7 Authors such as Nuno Simões Rodrigues (2007, 462 and 437) have argued that Sejanus—as the single Praetorian prefect since 17 CE—may have been behind this measure, as Philo of Alexandria accuses him of having hatched a nefarious plan to get rid of the Jews by leveraging his enormous influence on Tiberius (*In Flacum*, I, 1 and *Legatio ad Gaium*, 159–160). Nevertheless, if Sejanus was indeed the mastermind behind the *senatus consultum* of 19 CE, it makes no sense that Philo should fail to describe the episode, which he could have used to strengthen his accusations against the Praetorian prefect. Furthermore, the fact that Philo argued that Tiberius' policies towards the Jews were exemplary led him to use Sejanus as a scapegoat for anti-Jewish actions carried out under the emperor's reign. Cf. Castelli 2009, 76.

8 It is worth noting that it defies explanation why this alleged purge only affected young Jewish freedmen, and not the Jewish population as a whole.

9 Cf. Cassius Dio, LV, 31; LVI, 23; Velleius Paterculus, II, 111; Suetonius, *Aug.*, 25. See Cheesman 1914, 65–66; Applebaum 1971, 181–182; Baclay 1996, 299–300; Perani 2017, 117.

these Jewish soldiers were sent to Sardinia to fight brigands.[10] It is very likely that their surveillance tasks prioritized the protection of the most fertile areas (such as modern-day Barbagia) in order to secure supplies of grain.[11] Unsurprisingly, the most productive agricultural areas have yielded the greatest number of military inscriptions on the island.[12]

By assuming that the Jews drafted by Rome to be sent to Sardinia lacked Roman citizenship, some authors have concluded that these troops must have been auxiliaries.[13] However, the fact that they were clearly second-generation freedmen (*libertini*), as Tacitus expressly puts it, implies Roman citizenship, even if it may have been limited to *ius Latini Iuliani*,[14] which invalidates the above hypothesis. Indeed, it is very likely that the Jews who reached Rome as slaves in Pompey's time, or even earlier,[15] may have obtained freedom and even Roman citizenship by the reign of Augustus, according to Philo of Alexandria.[16] However, the fact that they were full Roman citizens would not support Samuel Rocca's thesis, according to which these soldiers—who, being 4,000, would have made up a standard legion of the period—were legionaries.[17] The impossibility of condemning citizens to *relegatio* without a trial[18] made *conscriptio* the only legal solution to expel citizens, who probably already made up the majority of Jews, from the capital of the Empire.[19] Augustus had set a precedent by drafting freedmen to protect the colonies near Illyricum and the left bank of the Rhine.[20] Even though it is true that, generally speaking, freedmen—the *plebs uernacula*—did not usually serve in the legions, the emperor had the power to grant them *restitutio natalium* through a decree and with the patron's con-

---

10 It was not the first time military units were dispatched to Sardinia to repress banditry: we know of at least two precedents within the same time frame (6 and 9 CE). See Le Bohec 1990, 23 and 97; cf. Castelli 2009, 67–80.

11 Castelli 2009, 74–75.

12 Le Bohec 1990, 99.

13 Castelli 2009, 79.

14 Johnston 2004, 44.

15 Smallwood 1976, 234–236.

16 Philo, *Legatio ad Gaium*, XXIII, 155–157. See Richardson 1998, 17–19.

17 Rocca 2010, 19. According to Birgit van der Lans (2015, 36), "the number of 4,000 freedmen or their descendants on which Josephus and Tacitus agree is hardly credible [...]."

18 Garnsey 1970, 116 and 119.

19 See Rocca 2010, 8, 16; Rodrigues 2007, 458–459. As Sherwin-White (1999, 324) aptly pointed out, even though the legal standing of manumitted freedmen and their children was not limited in any way, in practice, social prejudice impeded their access to certain prestigious titles and positions, such as those in the legions.

20 Suetonius, *Aug.*, 25. Other similar examples come from Cassius Dio, LV, 31 and LVI, 23, and Velleius Paterculus, II, 111. See Rocca 2010, 20.

sent. Thus, having become *ingenui*, they could join the legions as full Roman citizenship.[21] There is, however, no trace of this alleged legion exceptionally created by Tiberius,[22] and it is impossible to fit it within the rigid organization of the Roman army.[23] We cannot even assume that these freedmen may have been gathered into auxiliary units made up from Roman citizens, the so-called *cohortes uoluntariorum ciuium Romanorum*.[24] Had this been the case, the sources available to us would have pointed out the fact that the Jews who were drafted were citizens, rather than that they belonged to the *libertinum genus*.

We therefore cannot deem these Jews to have been soldiers in regular units of the Roman army, be they auxiliaries or legionaries. In fact, none of the sources display the terms commonly used at the time to refer to recruitment for these units (*dilectus, supplementum, conscribere*).[25] Nevertheless, our sources are unambiguous about their military role (*per speciem sacramenti*). Thus, in spite of its exceptional nature, this recruitment must have fit within a legal framework that defined the legal status that would govern it. It is obvious that the fact that they were freedmen did not prevent them from performing a military role, much like the *vigiles* (*libertini milites*) of the Augustan period.[26] Furthermore, Tacitus states that these Jews were of the ideal age (*idonea aetas*) for military service, and Suetonius mentions that only young Jews were recruited (*Iudaeorum iuuentus*). We could venture a hypothesis according to which, imitating the groups of *iuvenes* linked to the *principes iuuentutis*—Lucius and Caius, the adopted sons of Augustus—, that would go so far as to take on a paramilitary nature in the provinces,[27] there were "associations" of young Jewish Romans among which the "recruitment" drive ordered by the emperor may have taken place.[28]

---

21 Modestinus *Digest* 40.11.5. See Merrill 1919. According to Adrian Nicholas Sherwin-White (1999, 324), "freedmen certainly served in the Roman legions in times of crisis, in so far as their age permitted, and more frequently as marines, though they seem not to have been regularly enlisted at the annual."

22 See Rodríguez González 2001, which does not mention any legions stationed in Sardinia.

23 Edith Mary Smallwood (1976, 207) baselessly assumed that "all must in fact have been called up as auxiliaries, since there was no legion in Sardinia."

24 Glay 1972; Brunt 1974, 95 (= 1990, 194).

25 See Brunt 1974, 89–90 (= 1990, 188–189).

26 Sablayrolles 1996.

27 Ginestet 1991, 115–119; Le Bohec 2000, 32. According to Henry W. Pleket (1969, 286), "in general ancient youth-organizations, both Greek and Roman, seem to have focussed on physical, sometimes even downright military, training in combination with religious worship [...]."

28 As Philip A. Harland (2003, 201) has pointed out, in the Diaspora, though after the 1st

Despite the fact that our sources point out the anti-Jewish intent of the *senatus consultum*, I agree with Samuel Rocca when he states that "the Roman Senate's act, even in the context reported by Josephus as a period of tension between the Roman Jewish community and the Imperial authorities, cannot be seen as a discriminatory act towards a minority group, but rather as a harsh measure aimed at obtaining recruits at a time when Italic recruits were much less available than they had been in the past using the misbehaviour of some members of the community as an excuse for doing so."[29] The Jews seem to have been unable to oppose this measure on the grounds of its alleged illegality or by invoking the existence of hypothetical Jewish privileges.[30] Josephus states that "they penalized a good many of them, who refused to serve for fear of breaking the Jewish law".[31] However, they did not invoke a legal precedent that exempted them from military service in general, in contrast with the common practice of Jews from the eastern Diaspora.[32] Faced with a situation in which many of them were punished for opposing the draft, the fact that Josephus does not mention the exemption being invoked may lead us to think that it never applied to the western part of the Empire.

We do not know how long this Jewish military unit remained active in Sardinia or if it managed to accomplish the mission for which it had been raised, and neither do we know if there were any plans to dissolve the unit and return its members to Rome had it fulfilled its goal. We cannot prove the opinion of Edith Mary Smallwood, according to which they may have returned after the death of Sejanus (31 CE), if indeed he was responsible for their recruitment, or if they returned after Tiberius passed away (37 CE).[33] The fact is that, according to Cassius Dio, the Jewish population grew once again (αὖθις) in Claudius'

century C.E, "some Jews could participate in the activities of the gymnasium age-group associations, even forming age-group associations or joining those that already existed. Thus we find a reserved place for the association of Judaean youths (*Iouda[i]ōn neōterōn*) at Hypaipa (between Ephesus and Sardis); Jews among the young-men's organization (*ephēbes*) at Iasos (southeast of Miletos) [...]." However, there is also evidence of Jews among the ἔφηβοι at Cyrene in the late 1st century bce-early 1st century CE (Harland 2003, 301 note 14) and in Alexandria. See *BGU IV*, 1140, ll. 13 f.; Hunt and Edgar 1932–1934, II, 212 (= *P.Lond*, 1912), l. 94; Burr 2020, 65 = 1955, 18. On private associations in connection with the Jews of the Greek and Roman Diaspora, see now the works featured in Eckhardt 2019.

29 Rocca 2010, 18–19.

30 Cappelletti 2006, 61.

31 *Ant. Iud.*, XVIII, 81: [...] πλείστους δὲ ἐκόλασαν μὴ θέλοντας στρατεύεσθαι διὰ φυλακὴν τῶν πατρίων νόμων [...].

32 Rocca 2019, 228–229.

33 Smallwood 1976, 209; Laurenzi 2011, 18.

reign.[34] However, contrary to the opinion of some authors,[35] we cannot draw a direct relationship between the growth in the Jewish community in Rome and a hypothetical return of the Jewish soldiers who had been sent to Sardinia some twenty years earlier.

2 Jewish Troops in Roman Service

2.1 *Herod's Legacy*

After the death of Herod the Great (4 BCE), his three sons who survived him went to Rome in turn to request the entirety of the kingdom they felt they were entitled to from Augustus, who received them in the temple of Apollo and, having heard all parties, published his decision, which broadly confirmed the territorial distribution established in Herod's political testament. The eldest son, Archelaus (4 BCE–CE 6), was entrusted with half the kingdom: Judaea, with Jerusalem, Samaria, Idumaea, Caesarea, Sebaste and Joppe, although the cities of Gaza, Gadara and Hippus were separated from his domain and attached to the Roman province of Syria. He was stripped of the title of king, as Herod had intended, and given the lower rank of ethnarch (ἐθνάρχης) in its place. Antipas (4 BCE–CE 39), received the territories of Galilee and Peraea, with the title of tetrarch (τετράρχης). His brother Philip (4 BCE–CE 34), also as tetrarch, obtained the regions of Batanaea, Trachonitis, and Auranitis. Salome (4 BCE–CE 10), the sister of Herod the Great, also received a small portion of the inheritance: the cities of Jamnia, Azotus, Phasaelis, the palace of Ascalon and a hoard of 500,000 pieces of silver.[36]

Accused of behaving like a cruel tyrant by an embassy of Jews and Samaritans before the emperor, Archelaus was banished by Augustus to Vienne in Gaul in 6 CE, where he remained for a brief period of time before returning to the East, where he passed away. From that point onward and until 41 CE, his territory became a Roman province—subordinate to that of Syria under the generic name of Judaea—ruled by a governor of equestrian rank who had the title of *praefectus*. After the death of Philip (34 CE), who had no heirs, his territory fell under the province of Syria, although Caligula would later grant it to

34 Cassius Dio, LX, 6, 6.

35 Cf. Rodrigues 2007, 462–463.

36 Josephus, *Ant. Iud.*, XVII, 317–323; *Bell. Iud.*, II, 93–100. See Schürer 1973, 333; Gabba 1999, 127–128.

Agrippa I, the grandson of Herod the Great, with the title of king in 37 CE.[37] He would reinforce his power by obtaining the territory of the tetrarchy of Herod Antipas (Galilee and Peraea). Indeed, this tetrarch's position before Rome had been severely weakened with the total defeat (36 CE) in his confrontation with his former father-in-law, the Nabataean king Aretas, and the young Agrippa I, his nephew and brother-in-law, took advantage of this situation to accuse him of treason. Caligula exiled Antipas to *Lugdunum* in Gaul—where he would die in 39 CE—and the accuser Agrippa I was awarded his tetrarchy.[38]

The land of Palestine was divided into small Roman vassal kingdoms which the emperors gave and took away from the heirs of the various royal families, both for reasons of friendship and political expediency, turning them into autonomous kings. This was the case of Agrippa I, who became king of Judaea pursuant to an edict issued by Claudius (41–44 CE). Upon his death, his son Agrippa II kept the title but not the kingdom, which would go on to be governed (44–66 CE) by a Roman *procurator* who was once again subordinate to the governor (*legatus Augusti pro praetore*) of the imperial province of Syria.[39] Sometime after the death of his uncle, Herod of Chalcis (41–50 CE), Claudius gave him his kingdom in the Lebanon and the same charge of the Temple and right to appoint the High Priests that his uncle had enjoyed.[40] Scarcely a year after he formally took up the government of his kingdom (52 CE), the emperor granted him a much greater one in exchange for the small kingdom of Chalcis: the tetrarchy of Philip (Batanaea, Trachonitis and Gaulanitis), the tetrarchy of Lysanias (Abila), as well as the territory of Varus.[41] To all these possessions, Nero also added important parts of Galiee and Peraea.[42] In return, he remained ever-faithful to Rome, renouncing, as Emil Schürer put it, "even the small measure of independence which his father had sought to secure."[43] This can be seen in his military cooperation with Roman imperial power, when he provided, for instance, auxiliary troops for the Parthian campaign in 54 CE:

> [...] Nero ordered the young men recruited in the neighbouring provinces to be brought up to supplement the legions of the East, and for the legions themselves to be deployed closer to Armenia. He further ordered the

37 Josephus, *Bell. Iud.*, II, 181; *Ant. Iud.*, XVIII, 237. Gabba 1999, 130.
38 Josephus, *Ant. Iud.*, XVIII, 252; *Bell. Iud.*, II, 183. See Schürer 1973, 351–353; Gabba 1999, 132.
39 Gabba 1999, 142.
40 Josephus, *Ant. Iud.*, XX, 104; *Bell. Iud.*, II, 223.
41 Josephus, *Ant. Iud.*, XX, 138; *Bell. Iud.*, II, 247.
42 Josephus, *Ant. Iud.*, XX, 159; *Bell. Iud.*, II, 252. See Schürer 1973, 472; Gabba 1999, 144–145.
43 Schürer 1973, 474.

two veteran kings Agrippa and Antiochus [king of Commagene] to make ready forces for an offensive into Parthian territory [...].[44]

We do not know if the Jewish troops sent by Agrippa II "to supplement the legions" (*supplendis legionibus*) were able to maintain their religious customs intact as they fought side-by-side with gentile soldiers. This may have been the case as, despite his unconditional submission to Rome, Agrippa displayed much interest in ensuring that those customs were respected. His brothers-in-law Arizus of Emesa and Polemon of Cilicia were forced to be circumcised in order to be able to marry his sisters. Furthermore, the rabbinic tradition mentions certain questions concerning the Law which Agrippa's minister, or the king himself, addressed to the famous Rabbi Eliezer (ben Hyrcanus).[45] In this regard, we could assume that the Jewish king would not have allowed any military requirements to cause irreparable damage to his soldiers' Jewish religion. In fact, it seems that military service in support of a gentile army did not strip the Jewish troops sent into Parthian territory by king Agrippa II of their Jewishness.

In fact, this was but one of the many episodes that display the Herodian dynasty's traditional cooperation with the Romans, who were able to seize power in Judaea thanks to its support.[46] For instance, in 38 CE Antipas facilitated a meeting between the king of Parthia (Artabanus) and the then governor of Syria, L. Vitellius.[47] There is also no shortage of accounts of "collaboration" of Jewish troops in Roman military ventures. The Herodian dynasts made the Jewish military colonies founded by Herod the Great,[48] and which had become

44 Tacitus, *Ann.*, XIII, 7: [...] *Nero et iuventutem proximas per provincias quaesitam supplendis Orientis legionibus admovere legionesque ipsas propius Armeniam collocari iubet, duosque veteres reges Agrippam et Antiochum expedire copias, quis Parthorum fines ultro intrarent* [...]. See Schürer 1973, 474; Juster 1914, II, 270 note 2.

45 *Midrash Tanḥuma* (= *Yelammedenu*) [ca. 400 CE], *Lech Lecha* 20:4; BT, *Sukkah*, 27a:12; *Pesachim* 107b:2; *Kethuboth*, 17a:9–10. See Schürer 1973, 475.

46 Gracey 1986; Otzen 1990, 33–41; Richardson 1999, 24; Goodman 2007, 58–59, 376–382 (= 2008, 53–54, 359–365); Günther 2007, 89; Wilker 2007, 30; Rocca 2008, 52–58, 147–153.

47 Josephus *Ant. Iud.*, XVIII, 101. See Saddington 2009, 319. Even though we could assume it was the case, there is no evidence that Antipas took with him "un corps d'armée au secours de Vitellius" on that occasion, as was stated by Juster 1914, II, 270 note 2.

48 Josephus reports that Herod settled 3,000 Idumaeans in Trachonitis (*Ant. Iud.*, XVI, 285); 600 men of Zamaris in Batanaea (*Ant. Iud.*, XVII, 24), an unnumbered quantity of horsemen in Heshbon (*Ant. Iud.*, XV, 293–296) and Gaba (*Ant. Iud.*, XV, 294 and *Bell. Iud.*, III, 36), probably no more than 1,000 men; 6,000 colonists in Samaria (*Ant. Iud.*, XV, 296, and *Bell. Iud.*, I, 403), and 2,000 Idumaeans in Idumaea (*Bell. Iud.*, II, 55). See Isaac 1990, 327–331; Rocca 2008, 139 note 26 and 188–190.

a part of their armies, available to Roman generals on several occasions.[49] Nevertheless, we cannot ascertain whether these troops were integrated in the Roman military structure on a long-lasting or one-off basis, except in the well-known case of the military force of Zamaris.[50]

2.2 *The Troopers of Zamaris*

Ca. 20 BCE Herod the Great decided to establish a military colony in northern Judaea to act as a bulwark or shield against incursions by the inhabitants of Trachonitis (modern-day Sūr al-Leja, Southern Syria). He chose Zamaris, a famous Babylonian soldier of Jewish origin, to lead it; according to Flavius Josephus, at that time (ca. 10–7 BCE) he resided in neighbouring Antioch, next to Daphne of Syria, "with five hundred horsemen, all of them mounted archers, and a group of kinsmen amounting to a hundred men" (σὺν πεντακοσίοις ἱπποτοξόταις πᾶσι καὶ συγγενῶν πλήθει εἰς ἑκατὸν ἀνδρῶν).[51] Although the colony, located in the toparchy called Batanaea (modern-day Déraa), gradually grew with the arrival of contingents from diverse regions, Josephus states that "they were devoted to the ancestral customs of the Jews" (οἷς τὰ Ἰουδαίων θεραπεύεται πάτρια).[52] In order to safely host the entire Jewish Babylonian migration, Zamaris decided to erect a fortress—surrounded by several hamlets—which he named Bathyra:

> Being persuaded by this offer, the Babylonian went there to take possession of the land and built on it fortresses and a village, to which he gave the name of Bathyra. This man was a shield both to the inhabitants exposed to the Trachonites and to the Jews who came from Babylonia to sacrifice

---

49 Juster 1914, II, 270–272 note 2; Richardson 1999, 177–178, 259–260; Rocca 2008, 260–261; *Idem* 2019, 82–83. There are a number of Greek inscriptions of officers who belonged to these military colonies and to the army of Agrippa II: for instance, that of Herod son of Aumos, "stratopedarch of the cavalry of the colonials and infantry, also strategos to the great king" (*DMIPERP*, § 23), and Diomedes son of Chares, "prefect and general of Batanaea" (*DMIPERP*, § 30, § 31, § 32). See Momigliano 1967, 71–72; Gracey 1986, 320; Applebaum 1989b, 55–57, 61; Schwartz 1990, 117, note 35; Spaul 1994, 92–93; Sartre 2011; Zeichmann 2015, 184–187; *Idem* 2018, 5.

50 See Juster 1914, II, 270–272; Applebaum 1971, 181; *Idem* 1989.

51 Josephus, *Ant. Iud.*, XVII, 23–24. The use of Babylonian Jews as the core of a border military force was a characteristic feature of Herod's policies and those of his successors. In doing so, they fostered the introduction of Diaspora Jewish elements into the administration of Judaea, for they were unfettered by the patriotic loyalty that so influenced Jews born in the "motherland". See Applebaum 1971, 181; *Idem* 1979, 321; *Idem* 1989, 65; Isaac 1990, 329–330; Gichon and Vitale 1991, 253; Shatzman 1991, 175–180; Schoenfeld 2006, 117–118; Rocca 2008, 146, 189, and 260; Saddington 2009, 317; Zerbini 2016, 334.

52 Josephus, *Ant. Iud.*, XVII, 26.

> in Jerusalem; these he kept from being harmed by the brigandage of the Trachonites […].[53]

Thus, Bathyra became the military headquarters of the troopers of Zamaris,[54] thenceforth known as Bathyrans. Zamaris' son Jacimus inherited command over his father's troops and, after a period of great prosperity, he was in turn succeeded by his son Philip, who was a personal friend of Agrippa II and would go on to become the commander in chief of his entire army:

> When Zamaris the Babylonian, who had been acquired by Herod to take over the country, died after living a virtuous life, he left worthy sons. One of them was Jacimus, a man of shining valour, who organized the Babylonians under him into a body of horsemen; it was a troop of these men who served these kings as bodyguards. And when Jacimus died in old age he left a son named Philip, who had great manual strength and in respect of possessing other virtues could well bear comparison with any other man. For this reason faithful friendship and lasting goodwill were established between him and King Agrippa, and it was always he who trained any army that the king might maintain, and he would lead it wherever it had to march.[55]

Even though, upon the death of Agrippa I, all Palestine was taken over by Roman procurators (44–66 CE), the military colonies founded by Herod were allowed to remain in place. This is apparent in the fact that, according to Josephus, Agrippa II received both the Idumaean military colony and that of Zamaris in 53 CE, as Philip—the son of Jacimus—and his troops were then

53 Josephus, *Ant. Iud.*, XVII, 26: Τούτοις πεισθεὶς ὁ Βαβυλώνιος ἀφικνεῖται καὶ λαβὼν τὴν γῆν φρούρια ᾠκοδομήσατο καὶ κώμην, Βαρθύραν ὄνομα αὐτῇ θέμενος. πρόβλημά τε ἦν οὗτος ὁ ἀνὴρ καὶ τοῖς ἐγχωρίοις τὰ πρὸς τοὺς Τραχωνίτας καὶ Ἰουδαίων τοῖς ἐκ Βαβυλῶνος ἀφικνουμένοις κατὰ θυσίαν ἐπὶ Ἱεροσολύμων τοῦ μὴ λῃστείαις ὑπὸ τῶν Τραχωνιτῶν κακουργεῖσθαι […].

54 Applebaum 1989b, 52–53.

55 Josephus, *Ant. Iud.*, XVII, 29–31: Τελευτᾷ δὲ Ζάμαρις ὁ Βαβυλώνιος, ὃς ἐπὶ κτήσει τῆσδε τῆς χώρας Ἡρώδῃ προσποιεῖται, ζήσας τε μετὰ ἀρετῆς καὶ παῖδας λειπόμενος ἀγαθούς, Ἰάκειμον μέν, ὃς ἀνδρείᾳ γενόμενος ἐπιφανὴς ἱππεύειν συνεκρότησε τὸ ὑφ' αὑτῷ Βαβυλώνιον, καὶ ἴλη τῶνδε τῶν ἀνδρῶν ἐδορυφόρει τούσδε τοὺς βασιλέας. καὶ Ἰάκειμος δὲ ἐν γήρᾳ τελευτῶν Φίλιππον τὸν υἱὸν κατέλιπεν ἄνδρα κατὰ χεῖρας ἀγαθὸν καὶ τὰ ἄλλα ἀρετῇ χρῆσθαι παρ' ὁντινοῦν ἀξιόλογον. διόπερ φιλία τε πιστὴ καὶ εὔνοια ἀσφαλὴς αὐτῷ πρὸς Ἀγρίππαν γίνεται τὸν βασιλέα, στρατιάν τε ὁπόσην ὁ βασιλεὺς ἔτρεφεν οὗτος ἀσκῶν διετέλει καὶ ὅπῃ ἐξοδεύειν δεήσειεν ἡγούμενος. *Cf.* *DMIPERP*, § 293 (67 CE).

clearly at the Jewish king's service, and they subsequently served under the Romans in the Jewish War.[56]

At the time when Josephus was drafting his *Antiquitates*, the areas of Golan, Batanaea (the Biblical Bashan), Hauran and Trachonitis had been recently transferred to the Roman government (in the first year since Trajan's ascension, 97/98 CE),[57] as can be seen in a Greek inscription incised around the bust of a woman dressed in a tunic from Hauran (*Auranitis*, modern-day Jebel al-ʿArab) dated to 108 CE. It mentions an officer named Archieus, who had served as centurion under Agrippa II for eighteen years and under Trajan for ten.[58]

The *Archieus* inscription shows that Zamaris' troops were still active in 108 CE.[59] It is very likely that at least some of them carried on with their military activity independently for some time. The historian Flavius Arrianus[60] mentions troops known as "colonists" (οἱ ἀπὸ τῆς εἴλης ἧς ὄνομα Κολωνοί) who served in Cappadocia in 134 CE and who may have been derived from the ἱππεῖς Κολωνεῖται of Batanaea. This, according to Shim'on Applebaum,[61] allows us to identify them within the *Ala prima Augusta gemina Colonorum*. This unit, which was likely formed from two different contingents (on the one hand, the ἱππεῖς Κολωνεῖται and, on the other, the σπεῖρα Αὐγούστα), is epigraphically attested in Hadrian's time and in the 4th century, in Chiaca (between Satala and Melitene), under the command of the *Dux Armeniae*.[62] Cornelius Dexter, who commanded a fleet during the second Jewish War (132–135 CE), had previously been at the head of that unit.[63] In short, Zamaris' detachment had been entirely or partly transferred to a truly Roman force before 132 CE, possibly

---

56 Josephus, *Bell. Iud.*, II, 20, 1; IV, 1, 10; *Vit.*, 11, 35–36, 74. See also the inscription of the strategos Herod, son of Aumos, from Batanaea (modern-day Déraa) dated to 73 CE (*DMIPERP* § 23 = *AE* 1895.78 = Ewing 1895, 138). See Juster 1914, II, 272; Momigliano 1967, 71–72; Applebaum 1989b, 61; Isaac 1990, 330; Schwartz 1990, 117 note 35; Saddington 2009, 314, note 26; Sartre 2011, 6 (212).

57 *Ant. Iud.*, XVII, 22.

58 *IGLS XVI*, No. 1475 (= Seyrig 1965, 33 = *AE* 1966, 493): Ἀρχιεὺς ὁ ἐπὶ Ἀγρίππου βασιλέος γενόμενος κεντυρίων δεκαοκτὼ ἔτους καὶ ἐπὶ Τραιανοῦ στρατηγὸν δέκα. Now in the National Museum of Lebanon at Beirut. See Seyrig 1965, 31–34 with fig. 4; Applebaum 1989b, 61 and 63; Spaul, 1994, 92–93; Varon 2001, 274; Sartre 2011, 13 (220), No. 33 with fig. 7. According to Jean-Baptiste Yon and Julien Aliquot (2016, 220–221), the bust may represent the deceased wife of the official. The nominative case used suggests it is a commendation (Jacobson 2019, 160).

59 Applebaum 1989b, 63.

60 Flavius Arrianus, *De acie contra alanos* (or *Éktaxis*), I, 8.

61 Based, in turn, on Emil Ritterling (1902, 361–362).

62 *Not. Dig. Orientis*, XXXVIII, 21 (Neira Faleiro 2005, 283 = Seeck ³1983, 84).

63 Darmesteter 1880, 48.

during Trajan's reign.[64] Nevertheless, the troopers of Zamaris had already been fully romanized for a long time. The Romanization of the Herodian army and its ultimate absorption with the establishment of the Roman army in Judaea can be seen in the presence of Roman officers and cohort garrisons in Hauran and Batanaea (Bashan) at the time of the Agrippas, and in the later transfer of the troopers of Zamaris to a non-Jewish officer who was unrelated to the Zamarid clan.[65] It is likewise unsurprising that some of their members would later be transferred to certain posts within actual Roman units. Chares, the commander of the *Cohors Augusta* in Eithae, seems to have been a Jew and a relative of Philip, the former commander of the troopers of Zamaris.[66]

According to Shim'on Applebaum, the distinctive feature the troopers of Zamaris and of the unit of which they became the core is the fact that it is the last known Hellenistic-style κατοικία to have maintained its identity and military function after having been integrated in the Roman military structure. Its very existence shows that, despite the precarious political situation of the 1st century CE, a Jewish military unit could serve (albeit certainly under special conditions at first) in the imperial Roman army.[67]

2.3 *Collaborationism*

Some cases of desertion arose among Agrippa II's Jewish troops during the Jewish Revolt (66–70/74 CE).[68] Josephus even mentions the names of certain prominent figures that joined the Jewish insurgents.[69] Shortly after battle had been joined with the Roman army of Cestius Gallus near Jerusalem, Silas the Babylonian—apparently a distinguished officer from the Babylonian colony in Batanaea—abandoned his king to embrace the cause of the Jewish rebels.[70] Fighting by their side, he would ultimately meet his end in the siege of

64 See Applebaum 198b9, 63–64.

65 Mazzilli 2018, 48–53.

66 Josephus, *Vit.*, 177; *Bell. Iud.*, IV, 18. Χάρις (and Χαρῖνος) could have been regarded as an equivalent of the Hebrew name Ḥanan (חנן), "grace, favour", popular in Tanna'itic and 'Amora'ic times. See Reynolds and Tannenbaum 1987, 104–105; Applebaum 1989b, 57 and 62; Varon 2001, 274; Ilan 2008, 396.

67 Applebaum 1989b, 65. There is also evidence of Jewish military settlements in Asia (at least, in Galaunitis and Batanaea) at the service of the Roman army up to the early 2nd century CE (cf. the aforementioned Greek inscription of Archieus). See Seyrig 1965, 31–34; Applebaum 1974, 460; Sartre 2011, 13 (220).

68 Price 1992, 118–119.

69 On what follows, see Jacobson 2019, 77 and 161.

70 Josephus, *Bell. Iud.*, II, 520: "a deserter to the Jews from the army of King Agrippa" (αὐτομολήσας εἰς τοὺς Ἰουδαίους ἀπ' Ἀγρίππα τοῦ βασιλέως).

Ascalon.[71] In the defence of Jerusalem we also find Magassarus, formerly an attendant (θεράπων) at the court of Mariamme, Agrippa II's sister, who had switched sides to join the Zealots.[72] Josephus also tells us of a letter received in Galilee from an insolent cavalryman who had formerly served Agrippa II,[73] as well as of the death of one Iustus, a member of his bodyguard who had formerly held the same position with the king, during the defeat at Sephoris.[74] According to David Jacobson, "surely these defectors from Agrippa II's army are exceptions, but they signal a mixted reaction to the revolt by Jews serving in his army."[75] Nevertheless, there were many more Jews who defected to join the king's army. Important figures such as Iustus of Tiberias took refuge with the king when they were in dire straits.[76] Even entire cities, such as Tiberias itself, switched sides after having been initially hostile to Roman power.[77] Josephus also states that "the Tiberians were not the only traitors to your freedom, but so too were many of the most esteemed men in Galilee."[78] Conversely, the Galilean city of Sepphoris was loyal to King Agrippa II and to the Roman cause from the start.[79] There are even accounts of its inhabitants joining the Roman forces led by the tribunc (χιλιαρχοῦντος) Placidus: "now they offered a cordial welcome to the commander-in-chief [Vespasian], and promised him their active support against their countrymen (ὁμοφύλοι)."[80] This would not be, however, the only time Jews would fight their brethren during this war. Even though they eventually met a cruel end, the Jews of Scythopolis made a stand against the Jewish rebels to defend their city:

> Thus far the Jews had been faced with aliens only, but when they invaded Scythopolis they found their own nation in arms against them; for the Jews in this district ranged themselves on the side of the Scythopoli-

71 Josephus, *Bell. Iud.*, III, 11 and 19.
72 Josephus, *Bell. Iud.*, V, 474.
73 Josephus, *Vita*, 220.
74 Josephus, *Vita*, 397.
75 Jacobson 2019, 161.
76 Josephus, *Vita*, 354, 390–393.
77 Josephus, *Vita*, 381–385.
78 Josephus, *Vita*, 386: [...] οὐ γὰρ μόνοι Τιβεριεῖς προδόται τῆς ἐλευθερίας ἡμῶν γεγόνασιν, ἀλλὰ πολλοὶ καὶ τῶν ἐν Γαλιλαίᾳ δοκιμωτάτων.
79 Josephus, *Vita*, 30, 38, 104, 124, 346, 373.
80 Josephus, *Bell. Iud.*, III, 32: τότε γε μὴν φιλοφρόνως ἐκδεξάμενοι τὸν ἡγεμόνα [Οὐεσπασιανός] προθύμως σφᾶς αὐτοὺς ὑπέσχοντο κατὰ τῶν ὁμοφύλων συμμάχους. Cf. *Bell. Iud.*, III, 59, 110; *Vita*, 411. See Safrai 1970, 224–230.

> tans, and, regarding their own security as more important than the ties of blood, met their own countrymen in battle.[81]

After the ultimate destruction of Jerusalem, Josephus himself, who had earned the trust of Titus Caesar, "was sent with Cerealius and a thousand cavalry to a certain village called Thekoa, to ascertain whether the terrain was suitable for receiving a fenced camp."[82]

Even though a minority of the Jewish Royal Army of Agrippa II joined the Jewish rebels, the bulk of his troops remained loyal to their king in his resolute support towards of Rome.[83] No sooner had he received the sure threat of uprising from the embassy headed by Simon ben Ananias,[84] and before Cestius Gallus, the governor of Syria, mobilized his legions, the Jewish king sent Philip ben Jakimos, called a strategos (στρατηγός), perhaps highest-ranking army officer,[85] along with the hipparch (ἱππάρχης) Darius and 2.000 cavalry, including Bathyrans, to Jerusalem, where the Romans scarcely had a single cohort of some 500 men, to help the "peace party"[86] or, according to Josephus, to defend the people (ἐπαμυνοῦντας τῷ δήμῳ):

> Agrippa, on the other hand, equally solicitous for the rebels and for the nation against which they were rising in arms, anxious that the Romans should not lose the Jews nor the Jews their Temple and mother city, conscious, moreover, that he had nothing to gain from this disorder, dispatched to the aid of the citizens two thousand horse from Auranitis, Batanaea, and Trachonitis, under Darius, as cavalry commander, and Philip, son of Jakimus, as general.[87]

---

81 Josephus, *Bell. Iud.*, II, 466: Μέχρι μὲν δὴ τούτων Ἰουδαίοις πρὸς τὸ ἀλλόφυλον ἦσαν προσβολαί, κατατρέχοντες δὲ εἰς Σκυθόπολιν τοὺς παρ' ἐκείνοις Ἰουδαίους ἐπείρασαν πολεμίους· ταξάμενοι γὰρ μετὰ τῶν Σκυθοπολιτῶν [...]. Cf. *Vita*, 26. See Juster 1914, II, 278 note 6.

82 Josephus, *Vita*, 420: Πεμφθεὶς δ' ὑπὸ Τίτου Καίσαρος σὺν Κερεαλίῳ καὶ χιλίοις ἱππεῦσιν εἰς κώμην τινὰ Θεκῶαν λεγομένην προκατανοήσων, εἰ τόπος ἐπιτήδειός ἐστιν χάρακα δέξασθαι καὶ τῆς ἑαυτῶν ἀσφαλείας ἐν δευτέρῳ θέμενοι τὴν συγγένειαν ὁμόσε τοῖς ὁμοφύλοις ἐχώρουν [...].

83 Roth 2019, 90.

84 Josephus, *Bell. Iud.*, II, 418–419.

85 Price 1991, 79.

86 Price 1991, 79.

87 Josephus, *Bell. Iud.*, II, 421: Ἀγρίππας δὲ κηδόμενος ἐπίσης τῶν τε ἀφισταμένων καὶ πρὸς οὓς ὁ πόλεμος ἠγείρετο, βουλόμενός τε Ῥωμαίοις μὲν Ἰουδαίους σώζεσθαι, Ἰουδαίοις δὲ τὸ ἱερὸν καὶ τὴν μητρόπολιν, ἀλλ' οὐδ' ἑαυτῷ λυσιτελήσειν τὴν ταραχὴν ἐπιστάμενος, ἔπεμπεν τοὺς ἐπαμυνοῦντας τῷ δήμῳ δισχιλίους ἱππεῖς, Αὐρανίτας τε καὶ Βαταναίους καὶ Τραχωνίτας, ὑπὸ Δαρείῳ μὲν ἱππάρχῃ, στρατηγῷ δὲ τῷ Ἰακίμου Φιλίππῳ. Cf. *Ant. Iud.*, XVII, 29–31 and *Vita*, 46 ff. See Roth 2007, 415; *Idem* 2019, 90.

Although Josephus states that, through this initiative, Agrippa II intended to safeguard the city and its inhabitants, the Romans would not have trusted the Jewish king had he not clearly demonstrated that he was acting in their interests.[88] In any event, this initial attempt to crush the revolt was an absolute failure. Even though the rebels allowed Agrippa II's Jewish army to leave the city,[89] all the Roman soldiers were put to the sword except for their commander, Metilius, who, humiliated, was able to save his life.[90] The troops gathered by Cestius Gallus, the governor of Syria, would go on to be defeated by the rebels at Beth Horon. Philip, the commander-in-chief of the troopers of Zamaris, managed to escape with his life.[91] The Bathyrans then moved to Gamala, where Philip rejoined them.[92] They undoubtedly remained loyal to the Romans. In fact, the Jewish Royal Army, under the command of Modius Aequus, besieged Gamala in the Golan. They kept this strategic city blockaded for seven months, before being relieved by Roman forces. There can be no doubt that the bulk of the Herodian forces, including the Bathyrans, were there.[93] The prior collaboration of these troops undoubtedly allowed for the city to be finally taken on the tenth of November, 67 CE.[94]

Moreover Josephus, who had been appointed governor of Galilee by the general assembly of the Jerusalemites during the opening moves of the war,[95] tells us that another royal Jewish force was sent under a "decurion" (δεκαδάρχης) named Aebutius to engage the troops he himself commanded at Simonias, west of Nazareth.[96] It is said that not only had this decurion distinguished himself in the fight, but that he also "had shown the utmost gallantry and inflicted the severest losses on the Jews."[97]

---

88 Roth 2007, 415.

89 Josephus, *Bell. Iud.*, II, 437.

90 Josephus, *Bell. Iud.*, II, 453–454. According to Jonathan P. Roth (2019, 90), "[…] it is quite possible that this 'Roman' force was in fact a cohort of Samaritans, and that it was anti-Samaritan, not anti-Roman feeling that led to its massacre […]."

91 Josephus, *Vita*, 46.

92 Josephus, *Vita*, 47.

93 Roth 2019, 91.

94 Josephus, *Vita*, 114; *Bell. Iud.*, IV, 10 and 83. See Roth 2019, 91.

95 Josephus, *Vita*, 190, 193, 393. See Schwartz 2009, 35; Seward 2009, 58–76.

96 Early in the war, Gaba, near Galilee—about three kilometres north-west of Megiddo—served as the headquarters of the aforementioned decurion who commanded the troops in the Valley of Jezreel and provided him with auxiliaries (Josephus, *Vita*, 24, 115). See Isaac 1990, 329.

97 Josephus, *Bell. Iud.*, IV, 36: Πλεῖστοι μὲν οὖν Ῥωμαίων κατὰ ταύτην ἔπεσον τὴν μάχην, ἐν οἷς ὁ δεκαδάρχης Αἰβούτιος, ἀνὴρ οὐ μόνον ἐφ' ἧς ἔπεσε παρατάξεως, ἀλλὰ πανταχοῦ καὶ πρότερον γενναιότατος φανεὶς καὶ πλεῖστα κακὰ Ἰουδαίους ἐργασάμενος. See Roth 2007, 415.

According to Josephus, the royal bodyguard also joined the fray under its commander (ἡγεμών) Sulla, who managed to defeat him in Galilee despite being outnumbered (roughly 500 men facing an army of 5,000 rebels).[98]

Even though Jewish Royal forces are not explicitly mentioned in any of the operations leading up to the siege of Jerusalem, we may deduce that they were present from the fact that Josephus mentions that "the contingents from the allied kings, in greatly increased strength, were included in the Roman army."[99] In any event, there can be no doubt that Tiberius Julius Alexander,[100] a high-ranking official of Jewish extraction, was the *de facto* commander of the Roman forces:

> With these was the most tried of all his friends for loyalty and sagacity, Tiberius Alexander, hitherto in charge of Egypt in the interests of Titus and his father, and now deemed worthy to take command of these armies, because he had been the first to welcome the dynasty just arising and with splendid faith had attached himself to its fortunes while they were still uncertain. Pre-eminent moreover, through years and experience, as a counsellor in the exigencies of war, he now accompanied Titus.[101]

This character has often been deemed to be an "apostate" to the point of making him a "true" Roman citizen completely devoid of Jewishness.[102] Josephus states that he "did not stand by the practices of his people" (τοῖς γὰρ πατρίοις οὐκ ἐνέμεινεν οὗτος ἔθεσιν) despite the fact that his father Alexander Lysimachus, who had been alabarch in Alexandria, had precisely stood out for his religious devotion (πρὸς τὸν θεὸν εὐσεβείᾳ).[103] "But there is no knowledge of any formal act of apostasy on his part."[104] In fact, Josephus' comment might reflect a doctri-

98 Josephus, *Vita*, 398. Cf. *Vita*, 398. See Roth 2007, 416; *Idem* 2019, 92–93.

99 Josephus, *Bell. Iud.*, v, 42: [...] πρὸς οἷς αἵ τε τῶν βασιλέων συμμαχίαι πολὺ πλείους [...].

100 See now the renewed edition translated into Spanish from Burr's classic work 2020, 114–127 (orig. 1955, 67–82). *Cf.* DMIPERP, § 372, § 365, § 366, § 369, § 369, § 370, § 148 (*CPJ* II, 418; 418b; *OGIS I*, 586).

101 Josephus, *Bell. Iud.*, v, 45–46: Φίλων δὲ δοκιμώτατος εὔνοιάν τε καὶ σύνεσιν Τιβέριος Ἀλέξανδρος, πρότερον μὲν αὐτοῖς τὴν Αἴγυπτον διέπων, τότε δὲ τῶν στρατευμάτων ἄρχων, κριθεὶς ἄξιος ἐξ ὧν ἐδεξιώσατο πρῶτος ἐγειρομένην ἄρτι τὴν ἡγεμονίαν καὶ μετὰ πίστεως λαμπρᾶς ἐξ ἀδήλου τῇ τύχῃ προςέθετο, σύμβουλός γε μὴν ταῖς τοῦ πολέμου χρείαις ἡλικίᾳ τε προύχων καὶ κατ' ἐμπειρίαν εἵπετο. Cf. Tacitus, *Annals*, XV, 28, 3. See Williams 1998b, 95; Levick 1999, 47; Rajak 2005, 86; Schoenfeld 2006, 116–119; Olshanetsky 2018, 15.

102 According to Barbara Levick (1999, 27), he was "a renegade Jew." Cf. Pfeiffer 2008, 396.

103 Josephus, *Ant. Iud.*, XX, 100. Cf. *Ant. Iud.*, XVIII, 141. See Burr 2020, 67, 72 (orig. 1955, 20–21, 25).

104 See Roth ^{2}2007, 717 (= 1971, 1135). According to Tessa Rajak (2005, 86), he "was an apo-

nal dispute rather than an objective observation.[105] Jonathan P. Roth correctly pointed out, "in any case, Alexander's lack of piety did not make him any less Jewish."[106] It cannot be denied that, as the nephew of the famous philosopher Philo, he came from the most notable Jewish family in Egypt.[107] In fact, his Jewish compatriots never saw him as anything other than a Jew.[108]

In view of how events unfolded, the role played by the Jewish Royal forces in the suppression of the Jewish Revolt of 66–70/74 may have been somewhat limited, but the fact remains that it was very significant at certain junctures, especially at the beginning of the conflict.[109] We must not forget, for instance, that Agrippa II held onto Western Galilee, and although most of eastern Galilee joined in rebellion, he neutralized the important base at Gamala.[110] His unconditional support of the Roman cause never wavered, and Roman generals never went without his steadfast military support. That Jewish Royal forces were present throughout the war is proven by the fact that Vespasian removed them only after the end of the conflict.[111]

Without flatly denying the historiographical tradition that has aptly argued that the Jewish War of 66–70/74 was, in essence, a "War of Liberation" or an "Independence Struggle" tragically lost by valiant Jewish "nationalists" against a "arbitrary and cruel" Roman imperial State,[112] Jonathan P. Roth is not mistaken when he states that "it is perhaps better characterized as a civil war, in which one side, driven partly by class considerations, partly by religious fanaticism,

state—or perhaps merely non-observant Jew—for Josephus' Greek wording is not wholly transparent (*AJ* 20. 100)." Cf. Feldman 1993, 81; Étienne 2000, 125 and 139; Schimanowski 2007, 130 and 134.

105 Schimanowski 2007, 128: "[…] so könnte sich hinter der Bemerkung des Josephus auch eine grundlegende und wohlmöglich auch von persönlichem Interesse geprägte Kritik an den Landsleuten der Diaspora, vor allem gegenüber der Familie Philons, verbergen." Elsewhere, Josephus (*Ant. Iud.*, XVIII, 141) also stated that two of Herod the Great's grandchildren, the sons of Alexander, whom he had executed, renounced the observance (τὴν θεραπείαν ἐξέλιπεν) of what befits a Jew (τῶν Ἰουδαίοις ἐπιχωρίων) to take on the Greek traditions (μεταταξάμενοι πρὸς τὰ Ἕλλησι πάτρια).

106 Roth 2007, 410. Cf. Schimanowski 2007, 134. Even at the end of the war, Titus had qualms about burning the Temple, undoubtedly due to the influence of Tiberius Julius Alexander (who took pride of place among his generals), and indeed he unsuccessfully attempted to have the fire put out, though his orders were not heard in the din of battle (Josephus, *Bell. Iud.*, XX, 254–256). See Burr 2020, 121–122 (orig. 1955, 75–76); Schoenfeld 2006, 120.

107 Schimanowski 2007, 123–125; Roth 2019, 93.

108 Schimanowski 2007, 133.

109 Cf. Firpo 2009, 42.

110 Roth 2007, 415–416; cf. *Idem* 2019, 91.

111 Josephus, *Ant. Iud.*, XIX, 363. See Roth 2007, 417.

112 Shatzman 1999, 84.

attack not only foreigners but fellow Jews, and another of rulers and religious leaders trying to integrate their traditional culture into a modern world to save what they consider most important."[113] In his *Vita* Josephus describes, against the backdrop of the anti-Roman struggle, a near-permanent climate of civil war.[114] In fact, at some points, he dramatically expresses his fear that war could erupt in all its horrors if nothing was done to pacify the rivalry between Jewish factions.[115] In others, his deep sorrow when he realized that fratricidal conflict was inevitable becomes apparent:

> As I observed these things, I was very painfully distressed and ordered them to stop, reminding them that to treat compatriots in such ways was unholy.[116]

2.4 *A Hidden Jewish Identity*

In ancient sources, the notions of 'ethnicity' and 'identity' are usually joined together. "If an ethnic group—writes Michelle L. Mann—is defined as a group of people who can see themselves as different from other groups on the basis of perceived differences in socio-cultural practices, including religion, language, and/or ancestral origins, then ethnicity is all of the shared socio-cultural phenomena that enable an ethnic group to identify themselves as a collective entity." Thus, according to this definition, "ethnicity is one form of identity, since it is dependent on, and propagated by, comparative mechanisms that result from interaction with others."[117] Nevertheless, the semantic field of ethnic terminology does not always match the notion of socio-cultural and/or linguistic identity. Ethnonyms could acquire various meanings that could even change over time or adapt to usage in different areas. One same term could refer to the ethnic or geographical origins of a certain group or to their language. For the Romans, the term "Syrians" could both apply to inhabitants of Syria, whether Greek, Aramaic or even Latin speaking, and to people who spoke

113 Roth 2019, 93. According to Josephus (*Bell. Iud.*, II, 408–424), internal divisions between Jewish revolutionaries and the moderates who sought to maintain the *status quo* was as deep in Jerusalem as it was in the remaining cities in Palestine. Cf. Seward 2009, 128–129, 160.

114 Schwartz 2009, 35.

115 Josephus, *Vita*, 100, 265, 306.

116 Josephus, *Vita*, 377: Ταῦτ᾽ ἐγὼ θεασάμενος σφόδρα διετέθην ἀνιαρῶς καὶ παύεσθαι προσέταττον αὐτοῖς, ὑπομιμνήσκων ὅτι τοιαῦτα δρᾶν ὁμοφύλους οὐκ ἔστιν ὅσιον.

117 Mann 2008, 328.

Syrian, a common way of referring to Aramaic.[118] Thus, any Jew living in Syria or who spoke Aramaic might perfectly be described as a Syrian. For instance, Josephus refers to a Syrian centurion in Roman army called Gallus, who can clearly speak Aramaic, but whose "ethnicity" we cannot deduce directly from his name.[119]

Likewise, the use of religion as an ethnic marker associated to terms such as Ἰουδαῖοι or *Iudaei* is not without ambiguity.[120] Romans continued to use "Jews" as a blanket term for all different sorts of eastern monotheistic God-worshippers well into the 2nd century CE. For them, religious differences between Jews, Samaritans, and Christians were practically negligible. "Roman", in turn, might be a linguistic or a regional marker, but in the 1st century CE it was fundamentally a political appellation. There had been intense Roman colonization in the East for over a century. The continuous presence of at least half a dozen legions and the fact that an increasing number of inhabitants had Roman citizenship allowed the process of "Romanization" to advance at a much quicker pace than in other parts of the Empire. Even though many of these citizens, or their immediate ancestors, had emigrated from Rome, other individuals who had obtained or inherited this legal status were not Latin speakers. Paul and Flavius Josephus are well known examples.[121] As Jonathan P. Roth has argued, these individuals may, depending on the context, have been referred to as "Romans."[122]

There can be no doubt that, in pre-Christian times, a Hebrew name was a reliable indicator of Jewish identity, but we cannot rule out this possibility with Greek and Latin names either.[123] In fact, the interchangeable use of Semitic, Greek, and Latin names is characteristic of Jewish onomastic practices in Rome, Sicily, and other parts of the Roman Empire.[124] We are aware of Jewish individuals with Greek and Latin names, and even Hebrew names translated

---

118 Roth 2007, 409–410.

119 Josephus, *Bell. Iud.*, IV, 37. See Roth 2019, 80.

120 Williams 1997; *Idem* 2013, 24–28, 267–779.

121 Some of those Jewish Roman citizens may have received citizenship precisely because they fought for the Romans in Jewish military units. Although the manner in which Paul's father obtained Roman citizenship remains a mystery, it is possible that his father had served in a Herodian unit that was later added to the Roman arm as an auxiliary force. See Roth 2019, 86.

122 Roth 2019, 80.

123 We must also keep in mind that "qualities of individuals' names cannot predict their actions and beliefs during their lifetimes" (Stern 2013, 146). Cf. Mussies 1994, 274–275.

124 Rutgers 1995, 151; *Idem* 1998, 57, 145–151. According to this author, "in the Diaspora (except for the problematic evidence from Aphrodisias), the non-Jewish onomasticon appealed to Jews more than in Roman Palestine; and despite clear preferences for certain kinds of

into these languages or transcribed into their alphabets, both in rabbinic literature, New Testament writings and epigraphy.[125] Indeed, this is especially complicated with the military, when auxiliaries might adopt *tria nomina* to show of their newfound citizenship, complicating our sense of any earlier names they might have had or been born with.[126] Even if we were to ignore Jewish Roman citizens, we should not immediately assume that a Latin name equals Roman ethnicity, as it was fairly common for some Jews to use Latin names by the 1st century CE. This seems to have been the case, for instance, of the Jesus called Justus (Ἰησοῦς ὁ λεγόμενος Ἰοῦστος) mentioned by Paul in one of the epistles attributed to him.[127] Greek and Latin names make up the majority of inscriptions in the Jewish catacombs of Rome. Moreover, we have seen that some of the officers who served in the Herodian, or Jewish royal, army bore Latin names. Examples are Rufus the commander of the Royal Cavalry,[128] Volumnius, a military tribune,[129] and Aebutius, a decurion.[130] It is possible, as we have assumed until now, that these men were ethnic Romans serving in the Herodian army. Nevertheless, according to Jonathan P. Roth, "it is also possible, however, and in some cases more likely, that they are Jews with Latin names. While Herod and his successors certainly used Roman officers, we should not assume that there were no Jewish ones."[131]

---

names, this preference did not exclude the simultaneous use of names of different linguistic origin" (1995, 157). Cf. Williams 1998b, 95; *Eadem* 2013, 317–331, 363–381.

125 According to Tessa Rajak (1994, 240), "the interpretation of 'unJewish' material (a very different phenomenon from the plentiful inscriptions, amounting to some one hundred and twenty, which are not visibly Jewish and can only be defined as such by their context) is obviously a complicated matter." On the difficulties in detecting Jewish identity in Greco-Roman inscriptions that lack clearly Jewish symbology, see also Olshanetsky 2018, 11–12. Cf. van der Horst 1991, 130–131; Noy 1994; Rutgers 1998a, 73–95; Williams 2013, 263–264. According to the latter author, "in other long-established Diaspora communities, Jews invariably succumbed to the influence of local usage in the spheres of burial and memorialisation with the result that not only the forms of their tombs but the epitaphs also tended either to follow local conventions or to be close adaptations of them. That the Jews of Rome from very early on had both adopted and adapted Roman burial customs has long been recognised [...]" (2013, 192).

126 See *DMIPEPRP* § 295, which only lists the new name of the soldier (clearly taken after the emperor) and not his previous Semitic name, but all of whom have standard Roman names that give no sense as to their *éthnos*.

127 *The Letter of Paul to the Colossians*, 4:10–11. Cf. *Acts*, 4:11 and 18:7. See Williams 2013, 309–310, 329; Roth 2019, 86.

128 Josephus, *Bell. Iud.*, II, 52.

129 Josephus, *Bell. Iud.*, I, 535; *Ant. Iud.*, XVI, 332.

130 Josephus, *Bell. Iud.*, IV, 36.

131 Roth 2007, 410.

Josephus tells us that, during the rebellion that emerged in the wake of the death of Herod the Great (4 BCE) a general (στρατηγός) was sent into the Temple to negotiate with the rebels:

> Archelaus was becoming provoked at these things, but he withheld retaliation in view of the urgency surrounding his departure; he feared that after making an enemy of the rabble, he would then be detained by the commotion. He therefore tried by persuasion rather than by force to calm down the revolutionaries, and having secretly sent in the general, he kept appealing [to them] to desist.[132]

This general could be none other than Gratus, the commander-in-chief of Archelaus' armies,[133] but it is unthinkable that he was a Roman officer, as the presence of a gentile within the sacred enclosure of the Temple would only have angered the rebels, rather than appeased them.[134] Josephus, however, states that the revolutionaries were not appeased because their ill will towards Archelaus and his officers ruled out the possibility of negotiations.[135] It is, in fact, very likely that this *strategós* was the same allegedly Jewish commander who had arrested the forty young men for destroying the golden eagle while Herod was dying.[136] That would explain the hostility of the rebels to him, since they supported those who had smashed the eagle on the Temple gate.[137]

It is quite possible that the enigmatic Modius (*nomen*) Aequus (*cognomen*), a commander or quartermaster of the Jewish Royal Army, was in fact an ethnic Jew. The meaning of his name, 'Fair Measure,' might point to the fact that it was a nickname. And even Modius could also be the Latin version of a Jewish name such as Omri or Gomer. In any case, the possibility that it was a Jewish soldier who had adopted a Latin name could not be ruled out,[138] much like Sulla, the

132 Josephus, *Bell. Iud.*, II, 8: Πρὸς ἃ παρωξύνετο μὲν Ἀρχέλαος, ἐπεῖχε δὲ τὴν ἄμυναν ὑπὸ τῆς περὶ τὴν ἔξοδον ἐπείξεως, δεδοικὼς μή ποτε τὸ πλῆθος ἐκπολεμώσας κατασχεθείη τῷ κινήματι. Διὸ πειθοῖ μᾶλλον ἢ βίᾳ καταστέλλειν ἐπειρᾶτο τοὺς νεωτερίζοντας καὶ τὸν στρατηγὸν ὑποπέμψας παύσασθαι παρεκάλει. Cf. Josephus, *Ant. Iud.*, XVII, 209.

133 Josephus, *Bell. Iud.*, II, 58.

134 Roth 2007, 411.

135 Josephus, *Bell. Iud.*, II, 9.

136 Josephus, *Bell. Iud.*, I, 652.

137 Josephus, *Bell. Iud.*, II, 5–7.

138 It is certainly remarkable that two Roman soldiers from *Legio III Cyrenaica*, which participated in the Jewish War in 68 CE under the orders of Tiberius Julius Alexander (cf. Levick 1999, 47, 116; Rodríguez González 2001, I, 133), had these names: Modius Priscus (*P.Mich.* 7, 444) and Papirius Aequus (*CIL* 3, 6628).

officer at the head of Agrippa II's royal bodyguard who faced the troops of Josephus early in the war.[139] What seems an unlikely *cognomen* for a Roman of the late first century, but makes sense as a nickname or *cognomen* for a Jew.[140]

The case of the "decurion" (δεκαδάρχης) Aebutius who fought Josephus at Simonias is also curious. Jonathan P. Roth believes it is possible that he was yet another Jewish soldier with a Latin name, based on the fact that the troops he commanded were mostly "allies" (*sýmmachoi*) from Gaba, the site of a Herodian military settlement that was still active at the time.[141] In this case, however, we cannot ignore the fact that Josephus himself expressly states that Aebutius was an ethnic Roman decurion:

> In this engagement multitudes of Romans fell, including the decurion Aebutius, a man who had shown the utmost gallantry and inflicted the severest losses on the Jews, not only in the action in which he perished, but on all previous occasions.[142]

3 Material Evidence

3.1 *The Levant*

As we have seen, especially from the 1st century CE onwards, there were cases—as with Rufus, Volumnius or Aebutius, quoted by Josephus—, in which Latin names could, in fact, conceal a Jewish identity.[143] To this list we might add the soldier Titus Mucius Clemens, whose inscription (ca. 50–90 CE) on a marble tablet broken into five joining pieces, was unearthed near Dora (Bir el-Malik), Syria-Phoenice (figure 1). Here is the preserved part of the text:

> To Titus Mucius, son of Marcus of the tribe Clemens, *praefectus castrorum* of the great king Agrippa. *Adiutor* under Tiberius Alexander prefect of Egypt, prefect of *cohors I Lepidiana equitata*. *Adiutor* to [...] of Tiberius

139 Josephus, *Vita*, 399–406.
140 Roth 2007, 411; cf. *Idem* 2019, 91.
141 Isaac 1990, 328–329; Roth 2019, 91.
142 Josephus, *Bell. Iud.*, IV, 36: Πλεῖστοι μὲν οὖν Ῥωμαίων κατὰ ταύτην ἔπεσον τὴν μάχην, ἐν οἷς ὁ δεκαδάρχης Αἰβούτιος, ἀνὴρ οὐ μόνον ἐφ' ἧς ἔπεσε παρατάξεως, ἀλλὰ πανταχοῦ καὶ πρότερον γενναιότατος φανεὶς καὶ πλεῖστα κακὰ Ἰουδαίους ἐργασάμενος. There can also be no doubt that centurion Gallus, whom he mentions later on, was also an ethnic Roman (*Bell. Iud.*, IV, 37).
143 See pp. 31–32 and 59.

> Claudius [...], imperial *procurator* of [...]. Simonides and Z[...] sons of [patronym] to their [friend and benefactor,] an offering.[144]

The inscription is a dedication by two people (l. 9) to Titus Mucius Clemens, a Roman citizen of equestrian rank. The interpretation of the text has proved troublesome as the inscription is broken along the right side, requiring the text to be restored. In fact, Clemens' *cursus honorum* is only partly legible. Owing to its fragmentary conditions, scholars are divided as to whether his career is given in ascending or descending order. Altering the standard (i.e. early to late) order of the offices that made up the *cursus honorum*, Stephen Llewelyn argued that this Roman military man's last assigned position was precisely the first one on the inscription (*praefectus castrorum*), denying that he was an ascendant Jewish soldier.[145] This argument is based on the unproven assumption that *praefectus castrorum* was the highest office in his career.[146] Thus, "if the *cursus* is inverted, then the move is possibly explained by the appointment of Ti. Julius Alexander as chief-of-staff under Titus and the desire of Rome to place a trusted officer in an influential post in the forces of its ally."[147] This statement, however, is nothing but pure speculation that is entirely lacking in documentary support. In fact, we do not know of a single Roman general who was at the head of the Royal Armies of the Herodian dynasts. Josephus only mentions Jewish στρατηγοί like Philip ben Jakimos. There is no trace of Roman officers in inscriptions that feature Agrippa because the dedicator had been or was at the king's service. Two commanders of large task forces (*praefecti castrorum*, στρατοπεδάρ-

---

144 Martin 1983, 204 = Llewelyn 1998, 152 (*cf.* DMIPERP, §148; *CIIP II*, 2123): [Τί]τῳ Μουκίῳ Μάρ[κου υἱῷ ...] | [Κλ]ήμεντι, ἐπάρχῳ στ[ρατοῦ τοῦ] | βασιλέως μεγάλου Ἀγρίπ[πα, βοηθῷ] | Τιβερίου Ἀλεξάνδρου ἐπάρ[χου Αἰγύπτου] | ἐπάρχῳ σπείρης πρώτη[ς Λεπι-] | διανῆς ἱππικῆς, β[οηθῷ ...] | Τιβερίου Κλαυδίο[υ ...] | ἐπιτρόπου Σε[βαστοῦ Συρίας] | Σιμωνίδης καὶ Ζ[... οἱ] | υἱοὶ τῷ ἑαυτῶν [φίλῳ καὶ ευεργέτῃ] | χα[ριστήριον.] See Zeichmann 2018, 14.

145 Llewelyn 1987, 154–155.

146 Llewelyn 1987, 154. Cf. Martin 1983, 209. There is no evidence that, from the Roman point of view, the military rank held by the prefect (commander) of a mounted auxiliary cohort in the Roman army (*cohors I Lepidiana equitata*) was below that of *praefectus castrorum* of the army of a Roman-allied king, or to that of *adiutor* (= βοηθός, if the restoration above is correct) of a prefect of Egypt, Syria or any of the *procuraturae* in the Near East such as, for example, the *procuratura* Iamniae (Josephus, *Ant. Iud.*, XVIII, 158) put forward by Haensch (2014, 112–113). According to Michael Gracey (1986, 320), "it remains equally possible that he was a friend of the king's who began his service in Agrippa's army and proceeded with the king's recommendation into a post under Tiberius Julius Alexander in Egypt and thus into Roman service."

147 Llewelyn 1987, 155.

FIGURE 1
Greek funerary inscription for *Titus Mucius Clemens*, between 50 and 90 CE (?). Israel Museum, Jerusalem (inv. No. 3919)
SOURCE: *CIIP II*, 2123 (845)

χαι) are mentioned.[148] There are, however, examples of Jewish officers who not only completed their careers in the Roman army (such as Archieus), but who also reached an extremely high military rank (as with Tiberius Julius Alexander). Furthermore, taking into account that Clemens was assistant (*adiutor*) to the *praefectus Aegypti* from 66 onward—immediately after having served as commander of *cohors I Lepidiana equitata*—and concluded his career as prefect in king Agrippa II's army, probably during the final phase of the first Jewish war,[149] we would have to explain how he was *praefectus* of an auxiliary unit at a time when its existence is unattested.[150] There are therefore no significant reasons that prevent us from thinking that Clemens would have begun his career with a command in Agrippa's army and filled the posts on the inscription in the order in which they are listed.[151]

148 *AE* 1895, 78 = *IGR* III, 1144 = *OGIS* 425 (*cf. SEG* 48, 1945): στρατοπεδαρχήσαντι ἱππέων Κολωνειτῶν καὶ στρατιωτῶν; *AE* 1967, 525 = *SEG* 33, 1266 = 40, 1449 = *AE* 1987, 950: ἐπάρχωι στ[ρατευμάτος? τοῦ] βασιλέως μεγάλου Ἀγρίπ[πα]. For στρατοπεδάρχης as translation of *praefectus castrorum*, see for example: *IGR* III, 1432, and Dobson 1974, 415; Saddington 1995, 53–54. Several ἔπαρχοι, commanders of military units, two *centuriones*, commanders of subdivisions of such units, and two στρατηγοί are also recorded. See Haensch 2014, 108.

149 *CIIP II*, 846.

150 The earliest attestations for the disposition of the auxiliary unit are *CIL* XVI, 26 (Pannonia, 80 CE), *CIL* XVI, 45 (Moesia Inferior, 99 CE) and *CIL* XVI, 58 (Moesia Inferior, 114 CE). See Holder, 1980, 179, 189, 190 and 240.

151 Devijver 1976–1977, II, 581–582; Holder 1980, 78; Pflaum 1982, 130–132; Gracey 1986, 320. Cf. Kokkinos 1990, 128–136.

There seems to be no doubt as to the funerary nature of the headstone.[152] In fact, there is no honorary inscription for an officer of the army of Agrippa II, but several sepulchral inscriptions.[153] As results from the inscription, the military rank (*praefectus castrorum*) in the Herodian army leads us to think that Titus Mucius Clemens was a high-ranking Jewish official drawn from the local nobility.[154] It seems unlikely, however, that he obtained Roman citizenship (hence his *tria nomina*) after he joined the Roman army, as the difference in *praenomen* between Clemens and his father indicates he was not a first-generation citizen. We are unaware of the relationship between Simonides and Z[...], who seem to have been Hellenised Jews,[155] with Clemens. The editors of *CIIP II* restored the final two lines in a manner common in earlier scholarship: "sons, for their father, to remember him."[156] Nevertheless, were we to accept this possibility, it would be difficult to explain the "sons" apparent rejection of the *tria nomina*. In this regard, I believe Rudolf Haensch's interpretation to be more plausible: "Simonides and the other person could be clients or *vilici* of T. Mucius Clemens who had to erect his tomb because he died without leaving any close relatives behind."[157]

We know that, after the first Jewish War, some Jewish soldiers who had served in the army of Agrippa II, and even certain elite units that had been created by Herod the Great—like the troops of Zamaris—, became fully integrated in the Roman military structure, albeit under certain conditions. This is apparent in the aforementioned inscription of Archieus, "who was a centu-

152 Although the archaeological context in which this inscription was discovered in Bir el-Malik between Dor and 'Atlit is not clear, we can draw a connection with tombs excavated there in 1967 (see Kuhnen 1987 note 337). In this regard, it cannot be dismissed offhand that the tablet was once attached to a Jewish grave (Kuhnen 1987 note 338).

153 Haensch 2014, 113.

154 Gracey 1986, 320. On his Jewishness, see Schwartz 1984, 242 and *Idem* 1985, 296. According to Avi-Yonah (1966, 265), "The 'Herodians' must have been associated with like-minded gentiles who were favourably inclined towards Judaism: Titus Mucius Clemens might well have been one of those." Nevertheless, he does not manage to prove this hypothesis at any point in his paper.

155 Names formed from the root Σίμων (שמעון) are undeniably Jewish in origin. According to Tal Ilan (2008, 3), "the name Simon (Σίμων in Greek) is biblical (the second son of Jacob and one of the tribes of Israel) but is also a recognized, independent Greek name. Many Jews used this name, but whether they did so because they wanted a biblical name, or because they wanted a name that sounded both biblical and Greek, or they chose it as a Greek name with no idea of its biblical etymology, remains unknown. The name Simon is recorded 92 times in this corpus, making it the fourth most popular name. However, of these 92, only 38 are undeniably Jews." See also Ilan 2008, 165–168.

156 *CIIP II*, 845–846.

157 Haensch 2014, 113.

rion under King Agrippa eighteen years and was a *strategós* ten years under Trajan."[158] Other cases reveal Jews recruited to cover losses in certain auxiliary units in the Roman army. A military diploma from *Cohors I Ascalonitanorum equitata*[159] dated 19 March 144 CE displays the name of a probably Jewish soldier who, as a result of *honesta missio*, had been granted Roman citizenship after having served in that unit for twenty-five years. His name was Barcallipus son of Aggaeus, from Samosata (Samsat, Turkey), a city that was annexed to the Empire in 72 CE within the province of Syria.[160] Werner Eck and Andreas Pangerl note that the "bar" prefix is likely to be Semitic (as "Callippus" is a known Greek name) and "Aggaeus" is a rare name attested in the Hauran (*Auranitis*, modern-day Jebel al-ʿArab),[161] where Archieus hailed from and where Herod the Great had established one of his military colonies with Jewish soldiers from Babylonia (the troops of Zamaris).[162] According to these authors, the name Barcallipus was formed by combining two language elements, one being Greek in origin and the other stemming from an Aramaic root.[163]

Among the many graffiti discovered in the Jewish necropolis of *Besara*, or Beth Sheʿarim, situated by the modern town of Tivon in the southwestern Galilee in modern Israel, the one located in the first *arcosolium* (a kind of tomb excavated in the wall of the catacombs) of Room I from Hall C of Catacomb 4, roughly 1.5 meters above the ground and relatively close to the entrance, stands out in particular. Strokes depicting a standing human figure (ca. 15 cm high) portrayed as hairless, with deeply etched eyebrows, eyes, nose, and mouth, can be clearly seen. He is dressed in a military tunic with a belt and boots. He is holding a spear in both hands which he is pointing directly toward the tomb opening (figure 2).[164] Associated to the figure of this soldier due to its proxim-

158 *IGLS XVI*, No. 1475 (= Seyrig 1965, 33 = *AE* 1966, 493): Ἀρχιεὺς ὁ ἐπὶ Ἀγρίππου βασιλέος γενόμενος κεντυρίων δεκαοκτὼ ἔτους καὶ ἐπὶ Τραιανοῦ στρατηγὸν δέκα.

159 This auxiliary cohort of the Roman army was created in 62 CE by Cn. Domitius Corbulo in the context of a tumultuary levy to reinforce the border of the province of Syria on the river Euphrates against the Parthians. See *CIIP III*, 245–246; Jiménez de Furundarena 2018, 83.

160 *DMIPERP*, § 257 = *AE* 2015, 1903: [...] *A(nte) d(iem) XIIII k(alendas) Apr(iles) L(ucio) Aemilio Caro, Q(uinto) Egrilio Plariano co(n)s(ulibus). Coh(ortis) I Ascalonit(anorum) sag(ittariorum) cui praest P(ublius) Aurelius Aurelianus, ex pedite Barcallippo Aggaei f(ilio) Samo(satis)* [...]. See Eck and Pangerl 2015, 256; Jiménez de Furundarena 2018, 91; Zeichmann 2018, 6.

161 *SEG* VII, 1239. See Eck and Pangerl 2015, 256.

162 See MacAdam 2002, 50–51.

163 Eck and Pangerl 2015, 256: "So darf man wohl annehmen, dass in Barcallippus eine Verbindung eines griechischen und eines aramäischen Wortelements vorliegt."

164 Stern 2018, 107.

FIGURE 2 Beth She'arim necropolis. So-called "gladiator" graffiti. Hall C of Catacomb 4
SOURCE: STERN 2018, 108

ity, surrounding the first *arcosolium*, is a painted Greek epitaph dedicated to "Germanos [son] of Isak [Isakios] the Palmyrene."[165] It is undeniable that this name is much closer to a Roman, rather than a Jewish, cultural context. In fact, it could be said that "a man named Germanos, some imply, might act in more 'Roman' ways than did his father, whose parents had conferred to him an equivalent of the biblical name of Isaac."[166] Nevertheless, Germanos the Palmyrene

165 *IJO III*, Syr52 (82–83) = *CIJ II*, 1011: Γερμανὸς Ἰσακίου | Παλμυρηνοῦ. Cf. Applebaum 1971, 182; Olshanetsky 2018, 14; Stern 2018, 107.

166 Stern 2018, 127.

was undoubtedly Jewish, as he was buried in an exclusively Jewish necropolis.[167] Moreover, it is reasonable to accept that there is a direct relationship between his inscription and the nearby graffito of a man in military costume. According to David Noy and Hanswulf Bloedhorn, "human figures are often depicted in Palmyrene tombs, and Germanus' relatives appear to have brought a local practice with them to Beth She'arim."[168] The epitaph has been dated to the 3rd century CE (the Jewish necropolis of Beth She'arim began to receive burials, many of them in *kôkhim* ("oven tombs"), from the late second century through the early Byzantine period).[169] In fact, there are Palmyrene Jews buried in Beth She'arim from the late 2nd century to the 4th century.[170]

Nevertheless, if we are to deem *Germanos* to be a Jewish soldier (most certainly in the Roman army), we must first accept that the figure of a man in military costume that is closely related to his inscription may be plausibly identified as a soldier. Ever since it was discovered, this has been accepted as the most obvious and reasonable choice.[171] Nevertheless, a recent review of the meaning of graffiti similar to those discovered in other locations, such as Alexandria or Horvat Tel 'Eitun,[172] near Maresha/Eleutheropolis, has led Karen B. Stern to suggest a different interpretation of the armed figures from Beth She'arim.

167 "In the absence of explicit markers of pagan or Christian presence (and without hope of additional formal excavation in the foreseeable future inside the cemetery), moreover, the Beth She'arim necropolis conclusively contains the highest concentration of burials associated with Jewish populations in the Levant from the Roman and Byzantine periods." (Stern 2018, 86). In Rome, the vast majority of inscriptions that can be identified as Jewish feature simplified given names (*nomina singula*) from a semantic reference which replaces the *cognomen* as a *signum* or becomes an *agnomen*. The Jew Leontius provides a "casual" explanation in his own epitaph (*JIWE* 2, 104 = *CIJ I*, 32 = *CIL* VI, 39086a): *Amici, ego uos | hic exspecto, Leo nomine et signo Leontius* ("Friends, I wait for you here. My name is Leo and my *signum* Leontius"). See Perea Yébenes and González Salinero 2021, 356.

168 *IJO III*, 82–83.

169 Green 2008, 148; Zerbini 2016, 339. On the typology and chronology of the catacombs, see Weiss 2010, 213–231.

170 Rajak 1998b, 358. It would seem that Germanos was not the only Palmyrene Jew buried in the catacombs of Beth She'arim who was a civil servant of the imperial administration. A Greek inscription was found in the entrance to Hall K of Room III of the same catacomb 4 which mentions the *palatinus* Julianus: *IJO III*, 230 = *CIJ II*, 1006 = Schwabe and Lifshitz 1974, II, No. 61: Μημόριον Λεοντίου πατρὸς τοῦ ῥιββὶ Παρηγορίου καὶ Ἰουλιανοῦ παλατίνου ἀπὸ χρυσοχῶν (i.e. "Memorial of Leontius, father of Rabbi Paregorius and of Julianus the palatine from the goldsmiths"). See Stemberger 2000, 306; Roth-Gerson 2001, 208, No. LX; Schoenfeld 2006, 122; Rocca 2010, 28.

171 Mazar 1957, 126 (= *Idem* 1973, 182–183).

172 Dominant interpretations of these graffiti similarly posit that their images depict dead soldiers or gladiators buried in the caves. See Tzaferis 1982.

Just as in Hellenistic and Roman Alexandria, for example, reliefs and images of armed sentinels and composite animals such as sphinxes often flanked entrances to burial caves to deter tomb robbers,[173] visitors to Beth She'arim, or to Tel 'Eitun for that matter, may have etched graffito-figures with armor and weapons beside tombs for their "symbolic protection."[174] According to Karen B. Stern, "commemorators who visited these mortuary complexes might have scratched weapon-carrying figures at the borders of burial beds not to represent the images of the dead but to serve as sentinels to defend tombs and their contents from human or demonic intruders."[175]

There are, however, a number of serious issues with this theory. Not all figures are depicted bearing weapons or with a menacing demeanour. Many of them are not even near the entrance, nor do they look like sentinels. In fact, this same space (Hall C) also features etched images of gladiators—a *myrmillo* and a *retiarius*—locked in combat. It therefore seems that these figures had nothing to do with an alleged protective function, but rather, that these mortuary graffiti depict Jewish gladiators buried there.[176] The figure that concerns us, however, features a number of iconographic details (military belt and boots) that were clearly intended to portray it as a soldier. In fact, Karen B. Stern herself acknowledges that "images of soldiers and gladiators commonly appear in graffiti throughout the Roman world, from well-preserved cities like Pompeii in the Roman West to Dura-Europos and in Hatra in the East."[177] If, as she claims, "spatiality and location should play a more significant role in the interpretation of mortuary graffiti such as these,"[178] the absence of a single recurring pattern would rule out a direct correspondence in the particular interpretation of its meaning. Conversely, both the individual traits apparent in the figure of the soldier and the independent scenes depicting a gladiatorial fight point to the identification of the people buried there rather than towards images of an allegedly prophylactic or apotropaic nature.[179]

173 Empereur 1995; Corbelli 2006.

174 Stern 2013, 146–147; *Eadem* 2018, 127.

175 Stern 2013, 147.

176 Weis 2014, 166 and 211.

177 Stern 208, 127. See Maulucci Vivolo, 1993, 32, 33, 28 (= CIL IV, 1474), 30 (= CIL IV, 5215), 34, 37a and 37b (= CIL IV, 8055, 8056); Cooley and Cooley 2014, 88–89 (= 2004, 65–66) and *passim*.

178 Stern 208, 127.

179 A carved image from the Beth She'arim necropolis also shows a man in a frontal, upright position carrying a menorah on his head. Erwin R. Goodenough (1953, I, 92, 95) describes the figure as a Jewish soldier, but he maintains that it may have messianic or eschatological significance. See Hachlili 2001, 191 (a drawing of this graffito on p. 86). Its proximity to

3.2 *Egypt*

A wealth of information is available to us on the continued existence of old Jewish military units in Egypt during the High Empire. In fact, circumstances and the weight of tradition of Ptolemaic κατοικίαι led these military detachments or colonies along the river Nile to endure for some time (from the time of Augustus to the 2nd century at the earliest).[180] Flavius Josephus implies that, as with most Alexandrians, Jews suffered from the restriction to the supply of corn imposed by Germanicus in 19 CE due to the fact that they lacked Roman citizenship, but he likewise states that Rome had special faith in them when it came to the defence of the land:

> The administration of the corn supplies has, indeed, been withdrawn from them, as from the rest of the Alexandrians; but the most signal mark of the confidence reposed in them by the former kings, I mean the charge of the river and of the entire province (?), has been preserved to them by the emperors, who regarded them as not unworthy of such a trust.[181]

We are aware of the fact, for instance, that Jewish troops settled in "the land of Onias," who had been a part of the Ptolemaic army,[182] were gradually disbanded by the Romans, although there is clear evidence that the descendants of those soldiers remained in the area. It is quite possible that the last remains of the military colony were done away with in 73 CE, when the Jewish temple of Onias was finally closed and torn down by order of the Roman government.[183] In any event, a papyrus dated to the year 130 CE reveals that there were still Jewish military colonies in Roman service at the time. Its text mentions Jews who

---

the inscriptions of the Palmyrene Jewish group suggests that it may have been an epitaph relating to people from Palmyra. See Roth-Gerson 2001, 208, No. LX.

180 See Juster 1914, II, 274 note 3; Applebaum 1979, 202–220; Kasher 1985, 88–93; Otzen 1990, 54; Pucci ben Zeev 2005, 127. It is worth remembering, for instance, the Jewish detachment in Pelusium, in the eastern delta (Josephus, *Bell. Iud.*, I, 175). In his description of the Roman army in Egypt, Strabo (XVI, 4, 23) mentions that the Roman forces sent against the Arabs under the command of Aelius Gallus included some Jewish and Nabataean auxiliary units (*supra*, pp. 30–31).

181 Josephus, *Contr. Ap.*, II, 64: [...] *Nam amministratio tritici nihilo minus ab eis quam ab aliis Alexandrinis translata est; maximam ueroeis fidem olim a regibus datam conseruauerunt, id est fluminis custodiam totiusque custodiae, nequaquam his rebus indignos esse iudicantes.* The original Greek text of chapters 52 to 113 of book two of *Against Apion* has not survived, and only the Latin translation commissioned by Cassiodorus (6th century) is available. On river guarding—*potamophylacia*—as military service, see Premerstein 1903, 16.

182 See *supra*, pp. 17–18.

183 Josephus, *Bell. Iud.*, VII, 420–421 and 433 ff. See *CPJ I*, 52.

owned land in the Athribic district.[184] This land, which was undoubtedly confiscated from the Jews as a result of the crushing of their revolt in 115–117 CE, is called δημοσία (*ager publicus* in Latin) in the papyrus (lines 13, 16). It is very likely that these plots of land were previously categorized as γῆ κληρουχική. An inscription from Athribis reveals the military character of the local Jewish settlement in explicit terms with: οἱ ἐκ τῆς … [στρα]τιωτικῆς …[185]

As there were no laws banning Jews from serving in the imperial army, it would be unsurprising if many of these soldiers or their descendants went on to serve in properly Roman units. In this regard, an *ostracon* from Apollinopolis Magna (Edfu) dated to 116 CE points to the presence of Jewish individuals in the Roman army. Specifically, it is a receipt for the "Jewish tax" to be paid by a centurion named *Aninios* (a Greek transcription of the Hebrew name *Ḥananiah*, *Ḥanani* or *Ḥanina*)[186] for his slave *Thermauthos*:

> Thermauthos, slave of Ḥananiah the centurion paid the Jewish tax of three obols in year 19 of greatest Lord Trajan. Year 19 Pachon 23.[187]

This find is particularly surprising due to its exceptional nature, especially as it is dated at the time of the Jewish revolt in Egypt under Trajan's rule. It is worth pointing out, however, that this revolt only extended to the Edfu region after the date of the *óstracon* (May 18, 116), although it would shortly reach it.[188] We are unaware of whether or not the Jewish centurion was demobilized under said circumstances, as some authors have suggested based on the fact that the explicit acknowledgement of his Jewish identity clearly indicates that he was

---

184 *P.Oxy*. V, 500 = CPJ II, 448 (256).

185 Kasher 1985, 119.

186 The same name appears on another ostracon dated 107 CE (CPJ II, 212). It is even possible that it was the same man. According to Aryeh Kasher (1985, 79 note 18), "both ostraca were receipts for the payment of the 'jewish tax' by slaves bearing Egyptian names, and both were found in the Jewish quarter Delta."

187 *Ostrakon Edfu* 159 = CPJ II, 229 (135–136) = DMIPERP, § 183: Θερμαῦθ(ος) δοῦλ(ος) Ἀνιν(ίου) | κεντ(ουρίωνος) Ἰουδ(αικοῦ τελέσματος) ιθ' ∟ Τραιανοῦ | Ἀρίστου τοῦ κυρίου (τριώβολον). ∟ ιθ' Παχὼ(ν) κγ'. See CPJ I, 52; Applebaum, 1971, 181; Kasher 1985, 78–79; González Salinero 2003, 72–73; Zeichmann 2018, 6. Denying that Thermauthos was, in fact, a Jewish soldier serving in the Roman army, Mélèze-Modrzejewski (1995, 215) assumed that "he was simply a man whose father was named Kenturion." A 4th century inscription found in Jaffa of another Jewish soldier from Parembole (located on the Nile between Syene and Taphis) features a very similar epigraphic formula: DMIPERP, § 71 = CIJ II, 920 = JIGRE, 147 = CIIP III, 2240.

188 In fact, the disappearance of the Jewish community of Edfu after 116 CE was probably related to this revolt. See Alston 1995, 75.

under no circumstances willing to renounce his religion and that this may have been an obstacle to the fulfilment of his military obligations.[189] Nevertheless, this would not be the first time Jewish soldiers serving in the Roman army or fighting for Rome would face off against their brethren.

Another document from a slightly later period (late 2nd century CE) features a list of soldiers from one of the auxiliary units in the Roman army stationed in Egypt. According to its editor,[190] the list contains information on allegedly Jewish soldiers. The papyrus in question is in a poor state of conservation, and the only clues on the Jewish nature of these soldiers are the legible letters that form the beginning of the word *Ebra*, which could be reconstructed as *Ebra*[*eos*] (Ἑβραῖος), and the name *Abdior*, which is clearly Semitic in origin.[191] Another papyrus dated to a few years later (205 CE) displays similar features but poses more problems with regards to the Jewish identification of part of its contents.[192] It is a list of auxiliary soldiers whose names were not specifically Hebrew in origin (and who were therefore not necessarily Jewish), but rather, which stemmed from a common Semitic root, and they all seem to have hailed from Palmyra. Nevertheless, questions arise as to the identification of two of them, whose names may indeed be Jewish in origin: *Barichius* (ברך) and the *optio* Μαλωχῶς, i.e. *Malochus* (מלוכו).[193] In this case, there would be no contradiction between these soldiers being from Palmyra, like their comrades in arms, and Jewish at the same time.[194]

Beyond these clues, there are other accounts that have given rise to discussions as to whether they contain references to Jewish soldiers. The accounts in question are a series of papyri from the early 2nd century CE, penned by an officer named *Gaius Iulius Apollinarius*, who may have been a *principalis* who wished to become a *liberarius legionis*. He hailed from Karanis (in the Fayyum), although at the time (107 CE) he was stationed with his legion in Bostra (Syria).[195] Here is an excerpt from one of his letters:

189 Kasher, 1985, 78–79.

190 *Pap.Mich.*, VII, No. 448 (81–82) = *CPJ III*, 463.

191 Kasher 1985, 79–80; Varon 2001, 274.

192 *P.Oxy.* IV, 735 (227) = *CPJ III*, 465 (26). See Juster 1914, II, 274, note 3; *CPJ I*, 52; Kasher 1985, 152.

193 Juster 1914, II, 274; Negev 1991, 17, note 201; 39, note 637; Varon 2001, 275, note 44; Ilan 2008, 693 (on *Barichius*); Rocca 2010, 28.

194 *Cf. CPJ I*, 52 note 12.

195 On army veterans who returned to Karanis (the Fayyum) or decided to settle there, see Alston 1995, 123 ff.

> Iulius Apollinarius to Iulius Sabinus, his dearest father, very many greetings. Before all else, I pray for your good health, which is my wish, since I revere you next to the gods; but this has troubled me, that I have very often written to you through Saturninus the signifer, likewise through Iulianus the son of Longinus and through Dios, and not yet have you answered me concerning your health. [...] Greet Julia my lady sister, likewise Sarapias and my mother, my grandmother Sambathion, Thermouthis and her children, the father of Paccius, and all your colleagues individually, and those at home. I pray for your good health. The 10th year of Trajan, our lord, Phamenoth 30. || I am grateful to Volusius and Longinus Barbarus. You will tell the firm of Aphrodas, the son of the condiment dealer, that they enrolled me in the cohort at Bostra. It lies 8 days' journey from Petra and [...] || [...] soldier [...].[196]

As can be seen from the letters this soldier sent to his family over the years, there can be no doubt that he was fully integrated in the Roman army. Apollinarius' father Sabinus was also a soldier of *legio III Cyrenaica*.[197] His grandmother's name, *Sambathion*, has led some researchers to think he may have been a Jewish soldier, identifying it with *Sabbathaios* or *Sambathaios* ('born on Saturday').[198] Nevertheless, Victor A. Tcherikover expressed serious doubts

196 *DMIPERP*, § 167 (= *P.Mich.* VIII, 466; *CPJ III*, 486b; Campbell 1994, No. 36): Ἰ[ο]ύλιος Ἀπ[ολινάριος Ἰο]υλίῳ Σαβείνῳ τῷ | γλυκυ[τάτῳ πα]τρὶ πλεῖστα χαίρειν. | πρὸ τῶν ὅλ[ων εὔχομ]αί σε ἐρρῶσθαι, ὅ μοι εὐκτόν | ἐστιν, [ὅτι σέβομ]αί σε μετὰ τοὺς θεούς. τοῦ- | το δέ μ[οι ἠνώχ]λησεν ὅτι πλειστάκις | μου γρ[άψαντος διὰ] Σατουρνίνου τοῦ ση- | μεαφ[όρο]υ, ὁμ[ο]ίως διὰ Ἰου[λ]ιανοῦ τοῦ τοῦ | Λονγείν[ο]υ καὶ διὰ Δίου, καὶ οὔπω μοι ἀντέ- | γραψες περὶ τῆς σωτηρίας σου. ἀλλ' ὅμως | ἐρω[τηθ]εὶς ἀναγκαίως σχέθητι πρὸ πάν- | των [γρά]ψαι μοι περὶ τῆς σωτηρίας [ὑ]μῶν. [...] || Ἰούλιος Πρίσκος Ἀπολλινάριος [...]ίων [καὶ ο]ἱ κοντ[ουβε]ρ[ν]άριοι πάντες. ἀσπάζου Ἰουλίαν τὴν κυρίαν | μου ἀδελφήν, ὁμοίως Σαραπιάδα καὶ τ[ὴ]ν μητέραν, τὴν μάμαν Σαμβάθιον, Θερμοῦθιν καὶ τὰ | τέκνα αὐτῆς, τὸν πατέρα Πακκίου καὶ πάντας τοὺς κολλήγας σου κατ' ὄνομα καὶ τοὺς ἐν οἴκῳ. | ἐρρῶσθαί σε εὔχομαι. ἔτους ι' Τραιανοῦ τοῦ κυρίου Φαμενὼθ λ'. || εὐχαριστῶ Οὐολυσσίῳ καὶ Λονγείνῳ τῷ Βαρβάρῳ. μεταδώσις τοῖς παρὰ Ἀφοδᾶτος τοῦ τοῦ ἀρτυματοπώλο[υ] | ὅτι ἐστράτευσάν (με)ὶς χώ[ρ]την εἰς Βόστραν. κάθηται [π]ρὸ η' ἡμερῶν Πέτρας καὶ τ[...] || [...] | στρατ[ιώτου ...]. *Cf.* *DMIPERP*, § 168 = *P.Mich.* VIII, 465; *DMIPERP*, § 169 = *P.Mich.* VIII, 487; *DMIPERP*, § 170 = *P.Mich.* VIII, 486; *DMIPERP*, § 171 = *P.Mich.* IX, 562; *DMIPERP*, § 391 = *P.Mich.* VIII, 500; *DMIPERP*, § 392 = *P.Mich.* VIII, 501; *DMIPERP*, § 393 = *P.Mich.* VIII, 493; *DMIPERP*, § 394 = *P.Mich.* IX, 572; *DMIPERP*, § 395 = *P.Mich.* VIII, 498; *DMIPERP*, § 396 = *P.Mich.* VIII, 499; *DMIPERP*, § 397 = *P.Mich.* VIII, 496; *DMIPERP*, § 398 = *P.Mich.* VIII, 497; *DMIPERP*, § 399 = *P.Mich.* IX, 568–569. Cf. Speidel 1977, 691–694 (= *Idem* 1984, 233–236).

197 *DMIPERP*, § 388 = *P.Mich. IX*, 571; *DMIPERP*, § 389 = *P.Mich. VIII*, 485; *DMIPERP*, § 390 = *P.Mich. IX*, 549. A fuller discussion of the family's archive in *P.Mich.* IX, 5–8 and Vandorpe, Clarysse and Verreth 2015, 186–198.

198 On the truly Jewish nature of this name, see Ilan 2008, 42, 148–154 and *passim*.

on the Jewish nature of the name, which was common in Roman Egypt and especially in Karanis (the Fayyum) and Theadelphia, where the combination of Egyptian and Greek names within a single family was commonplace.[199] Thus, it would seem that Julius Apollinarius' parents were no exception to this. Even though it may have been a Hellenized Egyptian family[200] that achieved Roman citizenship through military service by some of its members,[201] we cannot rule out that one of its branches may have been Jewish in origin and that it may have been affected by the common process of cultural syncretism characteristic of Egyptian society. As Christopher B. Zeichmann points out, "while the name Sambathion does not prove with certainty his Jewish ancestry, it remains likely."[202]

In another papyrus, which was likewise found in Edfu and dated between 71 and 121 CE, there are doubts as to the identification of *Iesous*, son of *Papios*, as a soldier. The word δεκανός (soldier in command of ten men) is clearly present in the text, and, therefore, we might assume that said Jew may have served as a low-ranking officer (*decurio*) in a Roman auxiliary unit. Indeed, he was not a Roman citizen at the time and, according to other documents, which presumably refer to the same individual, he was subject to the tax called *laographía* (λαογραφία):[203]

> *Dekanos* Jesous son of Papios, Pesouris son of Jason, Dosarion son of Jesous, Sobtais son of Aulaios, Thedesis, a weaver, Sambathaios son of Simon.[204]

---

199 There is also the case of the guardsman Σωτᾶς ... τοῦ Σαμβατίωνος (CPJ III, 489e, dated to 155 CE), also from Theadelphia. For the discussion surrounding the name *Sambathious*, see CPJ I, 94–95. On the challenges of correctly identifying Jewish onomastics in Egypt, see Bagnall 1996, 276–277. Cf. Cornthwaite 2017.

200 Its Egyptian nature is also supported by references to genuinely Egyptian deities (Serapis and "other gods"). See Kasher 1985, 80.

201 In Egypt, as in other locations, it was common for veteran auxiliaries to be granted Roman citizenship, which was also applicable to their families, at the end of their service (see Alston 1995, 49).

202 DMIPERP, § 167, § 168, § 169, § 170, § 171. See also Zeichmann 2018, 6, and 17 note 16.

203 Kasher 1985, 79.

204 DMIPERP, § 172 = CPJ II, 405 (176): Δεκανὸς | Ἰησοῦς Παπίου, | Πεσοῦρις Ἰάσωνος, | Σοβτᾶις Αὐλαίου, | Θήδεσις γέρδις, Σαμβαθαῖ⟨ο⟩ς Σίμωνος. *Cf.* DMIPERP, § 173 = CPJ II, 298 (151); DMIPERP, § 174; DMIPERP, § 175 = CPJ II, 304 (153); DMIPERP, § 176 = CPJ II, 311 (154); DMIPERP, § 177; DMIPERP, § 178 = CPJ II, 321 (156); DMIPERP, § 179 = CPJ II, 220. See Worp 1986, 192–194; Zeichmann 2018, 6. The *laographía* tax, which affected males between the ages of 14 and 60 who did not hold Roman citizenship (women were exempted, as were the Greeks of Alexandria) was introduced in Egypt during the Augustan period (24/23 BCE). On this subject, see CPJ I, 61–62.

The editor of this papyrus does not specify whether he was a soldier or a police officer.[205] Aryeh Kasher seems to prefer the latter possibility, considering the fact that there are known cases of Jews working as policemen in other parts of Egypt.[206] However, we must keep in mind that the Roman army in Egypt always played a dual role: maintaining internal security by constantly policing the roads and preventing external attacks from Africa or the East through the Sinai border.[207] There is therefore no obstacle to believing that *Iesous* the Jew was a military officer.

3.3 *The Western Diaspora*

Among the rare epigraphic testimonies of Jewish soldiers in the imperial Roman army we must first point out *Matthaius*, son of the Syrian *Polaius*, a veteran of *legio I Adiutrix* who was granted Roman citizenship along with several of his comrades through a military diploma dated 23 December, 68 CE:

> Servius Galba, Emperor Caesar Augustus, chief priest, holding the authority of a tribune of the people, appointed consul for the second time, has given an honourable discharge and citizenship to those veterans who served in *legio I Adiutrix* whose names are written below, together with their children and descendants, and further has given them a grant of formal marriage with the wives which they had acquired at the time when citizenship was given to them, or if they were single at the time of the grant, with those wives whom they later took in marriage provided this is limited to one wife each. Dated to 11 days before the calends of January in the year when Gaius Bellius Natalis and Publius Cornelius Scipio were consuls. To Matthaius, the son of Polaius of Syria. Written down and copied from the tablet which has been set up at Rome in the Capitol beside the altar [...].[208]

---

205 *Pap.Tebt.*, I, 27 (105–114); *P.Oxy.*, II, 387 (314).

206 See Kasher 1985, 79.

207 Maxfield 2009; Perea Yébenes 2020, 278.

208 DMIPERP, § 293 = *CIL* III.2, 848, Dip. V = *CIL* XVI, 8 = *CIL* X, 771 = *AE* 1994, 387: *Ser(vius) Galba Imperator Caesar August(us), pontif(ex) maximus, tribunic(ia) potestate, co(n)s(ul) design(atus) II, vetaranis qui militaverunt in legione I Adiutrice, honestam missionem et civitatem dedit, quorum nomina subscripta sunt, ipsis liberis posterisque eorum, et conubium cum uxoribus, quas tunc habuissent, cum est civitas iis data, aut, siqui caelibes essent, cum iis, quas postea duxissent dumtaxat singuli singulas. a(nte) d(iem) XI k(alendas) Ianuar(ias) C(aio) Bellico Natale, P(ublio) Cornelio Scipione co(n)s(ulibus), Matthaio Polai f(ilio), Suro(s). Descript(um) et recogn(itum) ex tabula, quae fixa est Romae in Capitolio ad aram* [...]. See Applebaum 1970, 4; Buonopane 1994; Schoenfeld 2006, 118; Rocca 2010, 27;

Another military diploma from 5 April, 71 CE, features the name of Jewish soldier *M(arcus) Dama Surus Garasenus*.[209] Like *Matthaius*, he was a veteran who had served on a Roman war fleet (*classis*) based in the port of *Misenum*.[210] One of the soldiers acting as a witness, Lucius Cornelius Simon, from Caesarea Maritima, was likely Jewish too.[211] *Barsimsus Callisthenis*, from the same city, was a Jewish soldier who served in *Cohors I Vindelicorum*, stationed in the province Upper Dacia (Romania), from ca. 132 CE. He too was granted Roman citizenship through a military diploma dated 13 December 157 CE:

> [...] and all of whom are in the province Upper Dacia under the legate Statius Priscus and have been granted an honourable discharge having served 25 years or more and whose names have been written below. Dated to the calends of December in the year when Quintus Canusius Praenestinus and Gaius Lusius Sparsus were consuls. Granted to Barsimsus son of Callisthenes of Caesarea formerly pedes of *cohors I Vindelicorum milliaria* commanded by Lucius Versinius Aper, Hispello. Written down and copied from the bronze tablet which is in Rome on the wall behind the temple of the deified Augustus beside Minerva.[212]

Zeichmann 2018, 6. The same military diploma features the name of C(aius) Iulius Agrippa Apamma (from Apamea) "who—according to Shim'on Applebaum (1971, 184, note 16)—may well have been Jewish also."

209 DMIPERP, § 294 (= CIL XVI, 15; CIL X, 867; ILS I, 1990): [...] *gregali M(arco) Damae f(ilii), Suro Garaseno* [...] ("[...] to marine Marcus, the son of Dama, a Syrian Gerasene [...]"). See Kennedy 2004, 117. On the Jewish name *Dama*, see *JIWE I*, 7 (= *CIJ I*, 643): *L(ucius) Aiacius | P(ublii) l(ibertus) Dama | Iudaeus por|tor v(ivus) s(ibi) f(ecit)* ("Lucius Aiacius Dama, freedman of Publius, Jew, customs house worker, made (the tomb) while he was alive", from Aquileia, 1st century CE). See de Falco 2017, 267. Dama's name is also attested in the Talmud during the Tanna'itic period (*BT*, *Kiddushin* 31a; *Menachot* 99b; *Avodah Zarah* 23b, 24a, 27b; *Berakhot* 56b). Cf. Applebaum 1970, 5; Barclay 1996, 325–326.

210 Both *legio I Adiutrix* (*Classica Pia Fidelis Constans*) and *legio II Adiutrix* (*Pia Fidelis Constans Pannonica*) were recruited—the former in the year 68 and the latter in 70—from sailors who had until then served in the Roman war fleets stationed at *Misenum* and *Ravenna*, respectively. See Rodríguez González 2001, I, 32–34 and 73–75.

211 Applebaum 1970, 4–5; Varon 2001, 273–274. On the name of Simon (Σῖμον), see Ilan 2008, 343–344. According to Benjamin Isaac (1998, 105 = 1980–1981, 50): "He [Lucius Cornelius Simon] must have been a Caesarean Jew who fought in the Roman army during the First Jewish Revolt."

212 DMIPERP, § 296 (= CIL III.2, 882, Dip. XL = CIL XVI, 107): [...] *et sunt in Dacia Super(iore) sub Statio Prisco leg(ato), XXV plurib(us)ve stipend(is) emerit(is) dimissis honest(a) mission(e), quorum nomina subscripta sunt, civitat(em) Romanam, qui eorum non haberent, dedit et conubium cum uxorib(us), quas tunc habuissent, cum est civitas is data, aut cum is, quas postea duxiss(ent) dumtaxat singulis. Idib(us) Dec(embribus) Q(uinto) Canusio Praenestino, C(aio) Lusio Sparso co(n)s(olibus). coh(ortis) I Vindelicor(um) ∞, cui praest L(ucius)*

A new reading of an inscription allegedly found in Nazareth shows that a *C. Iulius Quartus* or *Quintus* served as a soldier in the *legio IV Flavia Felix*. This legion had been stationed in *Singidunum* (Belgrade) in *Moesia superior* since Domitian times. But, like many other legions, it took part in various campaigns with *vexillationes*, including to the East. According to Werner Eck and Dirk Koßmann, it seems possible that the *legio IV Flavia* or a *vexillatio* of it participated in the war against Bar Kochba.[213] Most significantly, this epitaph has the letter ל interspersed throughout, which is suggestive of Jewishness.

A mid-2nd century CE sepulchral inscription from *Aquincum* (Obuda, Hungary) mentions *Aelius Silvanus*, centurion in *Legio II Adiutrix*, referred to as a native of Syria-Palestine who was born in colonia *Aelia Capitolina* (Jerusalem).[214] However, in this case, there are no reasons to assume that he was a Jewish soldier, as Alexander Scheiber suggested.[215] Not even his place of birth could be put forward as evidence as, after the Bar Kochba war, *Aelia Capitolina* was an exclusively non-Jewish city.[216]

An inscription from the Jewish catacombs of Monteverde—or from a burial ground very close to it—, dated to the 2nd–3rd centuries CE, features the name of *L. Maecius Archon* with the title of *centurio alti ordinis* (figure 3):

> For L. Maecius L.f. Constantius, and for Maecia L.f. Lucianis, and for L. Maecius Victorinus, and for L. Maecia Sabbatis, his children, and for Julia Alexandria his spouse, L. Maecius L.f. Archon, centurion of high rank, (had this) made for the well-deserving ones.[217]

The above is the interpretation of the text put forward by Ivan di Stefano Manzella, which was in turn featured in Clauss-Slaby's *Epigraphik-Datenbank*

Versinius Aper, Hispell(o), ex pedite Barsimso Callistenis f(ilius), Caes(area). Descript(um) et recognit(um) ex tabula aerea qua Romae in muro post templ(um) divi Aug(usti) ad Minervam. See Applebaum 1970, 3; *Idem* 1971, 184 note 16; Benjamin, Spielmann and Stanciu 1990, 141; Gudea 1999–2000, 185; Varon 2001, 274; Schoenfeld 2006, 121; Rocca 2010, 28; Tibor 2016, 13; Olshanetsky 2018, 16; Zeichmann 2018, 5–6.

213 Eck and Koßmann 2015.

214 *IJO I*, App7, 331 (= *CIJ I*, 681b).

215 Scheiber 1983, 67.

216 Belayche 1999; Hadas-Lebel 2009, 190–191; Panayotov 2004, 62; *IJO I*, 331.

217 *JIWE* 2, 618 (cf. *CIL* VI, 39084 = *CIJ I*, 470 = Leon 1995, 333): *L(ucio) Maecio L(ucii filio) Constanṭiọ eṭ* | *Maeciae L(ucii filiae) Lucianidi eṭ* | *L(ucio) Maecio Victorino ẹṭ* | *L(uciae) Maeciae Sabbaṭidi filis* | *et Iul(iae) Alexandriae coniugi* || *fecit b(ene) m(erentibus) L(ucius) Maecius Ḷ(ucii filius)* | *Archon (centurio) alti ordinis*. Another photograph of the epigraph is available at: http://www.catacombsociety.org/characteristics-of-jewish-epitaphs-in-rome/nggallery/image/855.

FIGURE 3
Monteverde (Rome). Epitaph, 2nd–3rd Century (?). L. Maecius Archon. Museo Vaticano, lapidario ebraico ex-Lateranense (inv. n.º 30815)
PHOTO: CLAUSS-SLABY (EDCS-23102746)

(*EDCS*-12201343). Due to the fact that this author denied its Jewish nature, David Noy included it in Appendix 4 (Inscriptions not considered Jewish) of *JIWE 2*. Nevertheless, the fact that there are good reasons to accept this reading of the epigraph does not mean that we must necessarily deny its Jewish origin. In other words, the fact that *Lucius Maecius Archon* was a soldier does not preclude his possible Jewishness.

Even though the most common epigraphic abbreviation for *centuria* or *centurio* is an inverted "C" (Ↄ), in a number of cases, such as in the inscription at hand, it is S-shaped (ᶠ) and smaller and more angular than the letter "S" elsewhere in the inscription. Thus, in this case, the formula could be reconstructed as *centurio alti ordinis* ('centurion of high rank'). Ivan di Stefano Manzella acknowledges that there are no epigraphic parallels featuring the Latin phrase *altus ordo*, but it can be found in a passage of Seneca's *De constantia sapientis* in a military context: "The more honourable a man is by birth, reputation, and patrimony, the more heroically he should bear himself, remembering that the highest ranks stand in the front battle-line."[218] This parallel would allow us to establish an equivalence between *primus ordo* and *altus ordo*.[219] Likewise, we cannot forget that throughout the inscription, "F" for *filius* is omitted, a practice which Ivan di Stefano Manzella notes is known in military tombstones, especially at Carnuntum.[220]

I share Ivan di Stefano Manzella's opinion according to which, in this context, the word *archon* is a *cognomen* rather than a title. If we were to consider *archon* to be a title, regardless of how common it was among the Jewish community in Rome,[221] we would deprive *Lucius* (*praenomen*) *Maecius* (*nomen* or demonym) of the third part of his personal name, that is, the *cognomen*, which, in this case, would be reduced to an initial *I* (---). This would run con-

218 Seneca, *De constantia sapientis*, XIX, 3: [...] *Quo quisque honestior genere, fama, patrimonio est, hoc se fortius gerat, memor in prima acie altos ordines stare* [...].
219 di Stefano Manzella 1989, 110.
220 Cf. also *CIL* VII, 243 (*Eburacum*, Britannia). See di Stefano Manzella 1989, 106.
221 van der Horst 1991, 89–90; Leon 1995, 189–190; Ilan 2008, 41; Williams 2013, 144.

trary to both Roman and Jewish naming customs.[222] Moreover, there are no other cases in Jewish epigraphy, let alone in that from the catacomb of Monteverde,[223] which feature the formula *archon alti ordinis*, as Jean-Baptiste Frey claimed when he translated it as "archonte du Haut Conseil" (*CIJ I*, 470).[224] Conversely, it is common in both Jewish and pagan onomastics for names to reflect the profession, origin or title of the person in question. As it has been attested, for instance, with the name *Senior*, which is equivalent to πρεσβύτερος ('elder') in the synagogal hierarchy,[225] we could plausibly assume that *Archon* experienced a form of synecdoche, that is, a transfer of the signifier due to the contiguity of the signified and that, therefore, the title of *archon* had become a personal name.[226] This being so, we cannot understand why Ivan di Stefano Manzella ruled out the possibility that *L. Maecius Archon* was truly Jewish, suggesting he was a pagan. Nevertheless, as Margaret Williams aptly points out, "*Archon*, the most commonly attested synagogal title of all, even at Rome, was largely avoided by pagan religious societies."[227]

The same phenomenon could have taken place with the name of his daughter, *L. Maecia Sabbatis*. This *cognomen* is undeniably from a culturally Semitic cultural environment.[228] Like other names formed from the root *Sabbat*, which are amply attested in epigraphy (both Hebrew and Greek and Latin) and in papyrological sources,[229] this name undoubtedly derives from the Semitic word

222 di Stefano Manzella 1989, 108.

223 *JIWE* 2, 2 (= *CIJ I*, 317); *JIWE* 2, 4 (= *CIJ I*, 465); *JIWE* 2, 33 (= *CIJ I*, 291); *JIWE* 2, 69 (= *CIJ I*, 304); *JIWE* 2, 98 (= *CIJ I*, 316); *JIWE* 2, 100 (= *CIJ I*, 402); *JIWE* 2, 101 (= *CIJ I*, 325); *JIWE* 2, 110 (= *CIJ I*, 391); *JIWE* 2, 121 (= *CIJ I*, 324); *JIWE* 2, 124 (= *CIJ I*, 347); *JIWE* 2, 164 (= *CIJ I*, 337); *JIWE* 2, 165 (= *CIJ I*, 384); *JIWE* 2, 166 (= *CIJ I*, 390); *JIWE* 2, 167 (= *CIJ I*, 343); *JIWE* 2, 168 (= *CIJ I*, 442); *JIWE* 2, 193 (= *CIJ I*, 397); *JIWE* 2, 557 (= *CIJ I*, 380).

224 Cf. Leon 1995, 177–178; Williams 1998, 225: *Eadem* 2013, 138.

225 Iiro Kajanto (1965, 294) has six examples of the name in *CIL*. On the Senior elder see Burtchaell 2004, 237–240.

226 In Jewish sources generally *arch* always has a hierarchical function (Williams 1998, 226). Some Jewish names of Greek origin, such as *Archagatuhus*, *Archelaus*, *Archias*, *Archidamus*, *Archigenia* or *Archippus*, were formed from the root *arch*. See Ilan 2008, 228–229, 406.

227 Williams 1998a, 217. Cf. *Eadem* 2013, 120 note 55, 129.

228 Stern 2008, 267, note 27.

229 They are amply attested in Greco-Roman Egypt, both in epigraphy (*JIGRE*, 40, 58, 59, 60, 86, 93, 95, 96, 98, 106, 108) and papyri (*CPJ III*, 43–56 and 481–499). With regards to the Eastern Jewish Diaspora, *cf. IJO I*, App.21; *IJO III*, Syr.30. Aside from the city of Rome (*CIL* VI 4779, 5543, 10127, 11665, 16302, 20252, 21666, 22928, 23226, 23859, 24504, 26153, 34288, 37594, etc.) and Italy (Capua: *CIL* X, 3875, 4320, 4391; Ravenna: *CIL* XI 171; etc.), they are also recorded in Western Latin epigraphy: *JIWE I*, 68, 85, 126, 158.

Shabbat, which originally refers to the Jewish day of rest.[230] In fact, names formed from this word were very popular among Roman Jews,[231] but they were also used (albeit, logically, to a lesser extent) by some gentiles,[232] especially in the city of Rome, where it also achieved some level of popularity among slaves.[233] Names derived from the root *Shabbat* were also adopted, from the 2nd century CE onwards, by Christians close to Judaism, namely, Jewish Christians.[234]

In view of all of the above, considering the origins of the tombstone, it would be reasonable to assume that L. Maecius Archon was a centurion *primi ordinis* who served in a Roman legion whose name is unknown to us because, as Ivan di Stefano Manzella pointed out, the stone was preferentially given over to his family members. Due to the fact that it was not a personal inscription in which his *cursus honorum* would have been described in full detail, the party who commissioned it only had room to reflect his preeminent military status through a generic formula which, though novel for us, would have been intelligible to readers of that age.[235]

Rome has also yielded a funerary inscription dated to the 3rd century CE for a soldier called Rufinus (στρα|τευσάμενος ἀπὸ ταξέων, 'he campaigned in the ranks'). The painted Greek text was found in the hypogeum of a catacomb under Via Appia Pignatelli, not far from the Vigna Randanini Jewish cat-

230 Solin 1983, 640, note 120.

231 For the sake of example: Εὐσαββάτις (*JIWE II*, 108 = *CIJ I*, 379); Σαβάτυς (*JIWE II*, 47 = *CIJ I*, 395 and 294); *Sabbatius* (*JIWE II*, 7); Σαββάτις, masculino (*JIWE II*, 193 = *CIJ I*, 397; *JIWE II*, 356 = *CIJ I*, 156); *Sabbaṭis* (*JIWE* II, 618 = *CIJ I*, 470); Σαβάτιο (*JIWE II*, 220 = *CIJ I*, 263); Σαββατίος (*JIWE II*, 22 = *CIJ I*, 329); Σαββατία (*JIWE II*, 339 = *CIJ I*, 153); Σαβατίς (*JIWE II*, 19 = *CIJ I*, 394); Σαββατὶς, femenino (*JIWE II*, 269 = *CIJ I*, 157); Σαβατίδι (*JIWE II*, 244 = *CIJ I*, 155); Σαβάτης (*JIWE II*, 110 = *CIJ I*, 391); Σαββατ[...] (*JIWE II*, 157 = *CIJ I*, 373). See Ilan 2008, 152–154.

232 Solin, 1983, 645; Rodrigues, 2007, 191–192.

233 For instance, *CIL* VI 5041, 8494, 19163, 24887, 25519, 28209, 28267, 28361, 34518, 38438 (Most of which are from the 1st century CE). See Solin, 1983, 637–638, 645–646, 665, 679, 681 and 712. As shown by literary sources (Philo of Alexandria, *Legatio ad Gaium*, XXIII, 155; Tacitus, *Annals*, II, 85; Josephus, *Ant. Iud.*, XVII, 141; Martial, *Epigrammata*, 7, 35), we know that the Jewish community in Rome included both freemen (some of whom belonged to the social elite) and slaves. *Vid.* Jeffers 1998, 129 and 147; Ricci 2006, 93–96; Laurenzi 2011, 16–17.

234 In this regard, we know that Jewish Christian communities observed the Sabbath rest for a long time. On this subject, see Murray 2004, 57–59 and 85–86; Stern 2008, 133–134. In fact, Roman Christians giving up the "Sabbath" and adopting Sunday as their day of worship was part of a gradual process of differentiation between Christian religious identity and the Jewish creed (Bacchiocchi 1977, 169).

235 di Stefano Manzella 1989, 111.

acomb. Even though its discoverer, Nikolaus Müller (1857–1912),[236] identified the place as a Jewish burial ground, the impossibility to access this catacomb has prevented the German researcher's theory from being confirmed. Indeed, the minute description of the hypogeum he entered and the reproduction of the aforementioned inscription have failed to garner, with a few exceptions,[237] sufficient support from the scientific community.[238] Harry J. Leon merely noted that "whether this is a Jewish inscription depends on whether or not the catacomb was Jewish."[239] However, a detailed analysis of the information available on this catacomb and Rufinus' inscription should inevitably lead us to uphold their Jewish nature as the most plausible option.[240] There is, however, insufficient evidence to believe that, due to his name, the *grammateus* ('scribe') Castricius—to whom his wife Julia dedicated a headstone in the Vigna Randanini Jewish catacomb—, had been a soldier at any point of his life.[241] Even though his name may originally be related to the field of the military, this argument alone is insufficient to consider that possibility,[242] as Jonathan Roth suggests.[243]

Turning our gaze towards the provinces in the Danube, conquered lands that required constant military supervision, we discover the presence of Jewish soldiers among the Eastern troops that served among the Roman forces stationed there.[244] An inscription was found in Intercisa (modern-day Dunaújváros-Dunapentele), in Pannonia (Hungary), dedicated to emperor Severus Alexander (222–235 CE) and Julia Mamaea by the Jew Cosmius, a *praepositus stationis* (superintendent or commander of the road-station) (figure 4):

236 Müller 1886.
237 See Varon 2001, 275; Schoenfeld 2006, 123; Rodrigues 2007, 135.
238 Vismara 1986, 389–392; Rutgers 33 and 46.
239 Leon 1995, 274.
240 See the analysis in question in Appendix 2.
241 *JIWE II*, 266 = *CIJ I*, 221: *Castricius* | *grammateus.* | *Iulia coiiux* | *marito suo benemerenti* || *fecit.* | (open scroll?), i.e. "Castricius, grammateus. Julia his spouse (had this) made for her well-deserving husband." Cf. Leon 1995, 296 and fig. 58.
242 We do not even know if it was he who first bore that name or if he inherited it, as did his son. His name, which is identical to his father's, appears on the tombstone he dedicated to Julia, his mother (*JIWE II*, 267 = *CIJ I*, 124), within the same catacomb and very close to her husband's: τῇ ἰδίᾳ {μη} μητρὶ | Ἰυλίαι Καστρί|κις υἱὸς ἐποίη|σεν·ἐν εἰρήνῃ κοίτη | vac. σου. vac., i.e. "For his own mother Julia, Castricius her son (had this) made. In peace your resting-place." Cf. Leon 1995, 282.
243 Roth 2007, 418 note 41.
244 According to epigraphic remains from the province of Pannonia, Eastern Greek individuals made up approximately 10 % of the population. See Kovács 2004, 185–186.

FIGURE 4 Memorial tablet found in Intercisa (Dunaújváros-Dunapentele), Pannonia (Hungary), from 222–235 CE, preserved in the Hungarian National Museum, Budapest (inv. 154/1874, 2)

> To the Eternal God, for the welfare of our lord Severus Alexander Pius Felix Augustus, and of Julia Mamaea Augusta, mother of the Augustus, Cosmius willingly repaid his vow, superintendent (?) of the guard-post at Spondill …, from [i.e. a member of] the synagogue of the Jews.[245]

This inscription reveals to us that the military camp at Intercisa had a synagogue to which the dedicator belonged, and that it most likely provided religious services to the Jewish soldiers in camp.[246] Moreover, its construc-

---

245 *IJO I*, Pan3 (cf. CIL III.1, 3327; CIL III.Suppl. 10301; Erdélyi and Fülep 1954, 323, no. 329; *CIJ I*, 677; *RIU* V, 1991, 1051): *Deo Aeter|no pro sal(ute) d(omini) | n(ostri) Sev(eri) A[[ḷẹx̣an-]] | [[dri]] P(ii) F(elicis) Aug(usti) [[et Ịuḷ(iae)]] | [[Mamae]]ae Aug(ustae) mat(ris) Aug(usti) vot(um) | red(dit) l(ibens) Cosmius pr(aepositus?) | sta(tionis) Spondill(---) a synag(oga) |* (outside frame of tabula ansata): *I|u|de|ọr(um)*. See Fülep 1954, 258–259, 262, 268, 272; Applebaum 1971, 182; Fitz 1972, 175; Patai 1996, 22–23; Oppenheimer 2005a, 187; Schoenfeld 2006, 122; Kovács 2010, 161; Rocca 2010, 27; Tibor 2016, 15.

246 Applebaum 1971, 183.

tion implies the permanent presence of Jews among the troops settled in Intercisa, possibly since the conquest of the province during the principate of Augustus.[247] Perhaps that is why military commanders entrusted the protection of certain strategic locations to the Jews. One of them was undoubtedly the *statio* known as *Spondill* ... whose *praepositus*, at the time, was Cosmius.[248]

Three other epitaphs from Intercisa—likewise dated to the 3rd century CE—are claimed to show Jewish soldiers' names by Alexander Scheiber. The first one, *M. Aurelius Malchias*,[249] was *miles strator officii consularis* (a groom-soldier of the consular office) who served in *legio II Adiutrix*.[250] His father, *Mocur*,[251] was a *signifer* of *cohors milliaria Hemesenorum*.[252] *M. Aurelius Sallumas*,[253] a veteran *ex tesserario* who passed away at the age of 82,[254] and *Germanius Valens*,[255] whose wife was called *Aurelia Baracha*, also served in the same unit.[256] Heikki Solin, however, believed that these soldiers were, in fact, Syrian.[257] It is true that these names can be considered either Jewish or Syrian.[258] Nonetheless, considering that the Syrian city of Emesa (*Augusta Libanensis*), modern-day Homs, had a large Jewish community,[259] we cannot rule

247 Patai 1996, 24.

248 Cf. *CTh.*, 7.4.12. On *statores* Ulpianus *Digest* 4.6.10. We must keep in mind that, in a military context, *stationes* were barracks meant to house small groups of soldiers whose mission was to patrol and monitor certain strategically-important areas. The control they exerted over their assigned territories made *stationes* little beacons of Roman power. See Nelis-Clément 2006; France and Nelis-Cléments 2014.

249 On the Biblical origins of this name, see Scheiber 1983, 35.

250 *IJO I*, 7–8 = Erdélyi and Fülep 1954, 280, No. 17 = *RIU* V, 1189. See Fülep 1954, 250–251; Scheiber 1983, 33–35; Fitz 1972, 175; Varon 2001, 275; Rocca 2010, 28.

251 On this name, see Ilan 2008, 699.

252 See Fülep 1954, 250–251; Fitz 1972, 153; Rocca 2010, 28.

253 On this name, see Scheiber 1983, 42; Ilan 2008, 160.

254 *IJO I*, 8 = Erdélyi and Fülep 1954, 198–199, No. 133 = *RIU* V, 1202. See Fülep 1954, 200, 254, 259; Fitz 1972, 130; Rocca 2010, 28.

255 *IJO I*, 7 = Erdélyi and Fülep 1954, 281, No. 19. On the name *Germanus*/*Germanius* in a Jewish context, see Leon 1928, 207; Ilan 2008, 498.

256 On the undeniably Jewish nature of this name (ברכה), see Fülep 1954, 251; Scheiber 1983, 39; Ilan 2008, 176. According to Varon (2001, 275), *Aurelius Eliaseir ... natione Pannonius*, an instructor in *cohors VIII praetoria* (*CIL* VI, 2697 = Dobó, 1975, 15, No. 5), may also have been a Jew. On the name אלעזר (*Eleazar*), see Ilan 2008, 77–79.

257 Solin 1983. Cf. Kovács 2010, 175.

258 Cf. Tibor 2016, 12–13.

259 According to Walter Ameling (2009, 220), "it is estimated that 12 per cent of the population of Edessa was Jewish, although this figure is surely much too high." Cf. Drijvers 1992, 138.

out that some soldiers in *cohors milliaria Hemesenorum* were Syrian Jews.[260] These two identities are not necessarily mutually exclusive. The soldier *Aurelius Damas* is a clear example of this.[261] Whereas Ferenc Fülep believed he was clearly Semitic, László Barkóczi did not rule out the possibility that he was Syrian in origin.[262] It is true that this name is attested both in the Antioch region and around Emesa,[263] but there are also undisputable cases of its use by Jews: *Lucius Aiacius Dama* (*Iudaeus*)[264] or *Marcus Dama Surus Garasenus*.[265] The latter, moreover, clearly highlights the bearer's dual identity as a Syrian and a Jew.

The Latin inscription of the Jew *Joses*, from Oescus (modern-day Gigen) on the Danube, in Moesia (Bulgaria), may be from a slightly later date:[266]

> Joses, archisynagogos (?) and *principalis*, son of Maximinus the Pannonian, for himself and Cyria his spouse, dedicated the memorial while he was alive.[267]

From mid-first century CE onward, Oescus was the home-base of *Legio V Macedonica*.[268] Although the military rank of *principalis*—a non-commissioned officer—held by Joses in this legion is insufficiently attested in inscriptions,[269] it is described as an outstanding officer in a legal text by Ulpian collected in

---

260 Roux 2020, 29–30. It is quite possible that the stationing by Marcus Aurelius of an Emesene military unit at Intercisa, after the destruction of the fort in the Marcomannic War, led to the recruitment of Emesene Jews. See *IJO I*, 7.

261 Erdélyi and Fülep 1954, 324, No. 340. Cf. Solin 1983, 642, 649.

262 Fülep 1954, 253, 259; Barkóczi 1964, 310.

263 Fitz 1972, 147.

264 *JIWE* I, 7 (= *CIJ I*, 643). See de Falco 2017, 267.

265 *DMIPERP*, § 294 (= *CIL* XVI, 15; *CIL* X, 867; *ILS* I, 1990). See Zeichmann 2018, 5–6.

266 According to Alexander Panayotov (2004, 61), "a possible explanation of the way in which the Jews settled in Moesia is the pattern followed by other immigrants from the eastern provinces of the Roman Empire. Natives from the East are usually attested in the Danube provinces as conscripts in the army units stationed there or as veterans." Cf. Mócsy 1974, 228–230.

267 *IJO I*, Moes1 = *ILBulg*, 67 (cf. *CIJ I*, 681; Scheiber 1983, 57–61; Kovács 2010, 162): [----------] | *Ioses arcisina(gogus?)* | *et principales (hedera)* | *filius Maximini* | *Pannoni sibi et* | *Qyriae coiugi* | *sui vivo suo me|moria dedica|* (palm branch) *vit.* (hedera). See Panayotov 2004, 58–60; Schoenfeld 2006, 122; Olshanetsky 2018, 23. The name Ἰωσῆς (Joses, Joseph) was very popular among Jews, both in Palestine and in the Diaspora. It appears in the Balkans, for instance, in the inscriptions of Beroia (*JIGRE*, 143), Argos (*CIJ I*, 719), Corone (*CIL I.2*, 721c) and Jajce (*ILJug*, 149). See Panayotov 2004, 58.

268 Rodríguez González 2001, I, 195.

269 *CIL* V 1693, *CIL* X, 1608.

the *Digest*.[270] His rank was consistent with his role as *archisynagogus* or head of the local Jewish community.[271]

Finally, it is worth pointing out the existence of a gold amulet (gold *lamella*) found on the southwest side of the military garrison of *Segontium* (Caernarvon, Wales) with magic signs and Jewish liturgical fragments in Greek. Letterforms suggest that this tablet belongs to the earliest period of the site (ca. 75–140 CE).[272] Assuming that the amulet—which was intended to provide protection from death itself and from the woes to be encountered in the Afterlife—, was not stolen or taken as war booty, its owner may very well have been a Jewish soldier who was then serving in one of the Roman army units stationed in Wales.[273]

4 Dura-Europos

The city of Dura-Europos, located on the border of the Roman and Sasanian Empires, was home to a major Roman garrison in the third century CE. The varied cultural origins of the soldiers stationed there were largely reflected in their places of worship, to which they were closely attached owing to their religious beliefs. Thus, we can see that Roman soldiers from two Syrian legions attended rites held in the Temple of Bel, although we also know that they built the second phase of the *mithraeum* in 209–211.[274] Some of the inscriptions discovered in this *mithraeum* name a body of Palmyrene archers, which existed in the earliest stages of Roman occupation.[275] This did not prevent the Palmyrene cohort from using the Temple of Artemis Azzanathkona as its headquarters:[276] a number of images of soldiers making sacrifices to their gods have been found

270 Macer *Digest* 49.16.13.4. See *IJO I*, 33; Panayotov 2004, 60; Schoenfeld 2006, 125; Tibor 2016, 13.

271 Believing it inconceivable for an *archisynagogus* to also be an officer in the army, Péter Kovács (2010, 172) unconvincingly argued that Joses was a civilian *principalis*, i.e. a leader of his town.

272 Kotansky 1994, 1–12.

273 Varon 2001, 274. The fact that the amulet was likely from the military cemetery (Kotansky 1994, 3–4) and its protective role against death may lead us to think that its owner—presumably a Jew—was, in fact, a soldier. Indeed, like many soldiers, he may have been superstitious due to his constant exposure to the risk of death and therefore had it made as a personal phylactery.

274 The cults of Mithras and Jupiter Dolichenus had well-known ties to the Roman army. See Nock 1952, 200, 222; Pollard 1996, 221–223.

275 Welles *et alii* 1959, 24–25.

276 Welles *et alii* 1959, 25.

FIGURE 5
Dura-Europos Synagogue. Detail of the frescoes on the W wall: Moses parting the Red Sea and the crossing of the Twelve Tribes of Israel. The Israelites, with their long trousers and oval shields, bear a strong resemblance to the Roman soldiers from Dura, especially considering that each tribe is headed by a standard-bearer carrying a *vexillum*.

inside.[277] Christian soldiers also began to be present from the early 3rd century CE onward.[278]

In and of itself, the existence of a synagogue in Dura-Europos demonstrates the presence of Jews in the city since the 2nd century CE at the latest.[279] Nevertheless, we lack direct information on their possible presence among the Roman troops serving on the remote Eastern border with the Sasanian Empire.[280] According to the propaganda inscription from the *Kaabah of Zoroaster* or *Kaʿab-Ye Zardošt* (ŠKZ),[281] the Sasanian king Shapur I (241–272 CE)—who conquered many cities in Mesopotamia, Syria, and Cappadocia, including, first and foremost, Dura-Europos (256 CE)—, encountered a Roman force of seventy-thousand men that included soldiers from Judaea.[282] Even though this is nothing but a sweeping statement, it is possible that some of the Jews living in Dura were active participants in the Roman army. We must keep in mind that block L7 "was not only the location of the synagogue precinct, but also, the place where a large part of Dura's Jewish community lived, as well as an area where soldiers were billeted."[283] In this regard, some scholars have drawn

277 Pollard 1996, 223. Lucinda Dirven (1999, 184) writes that "the Palmyrene soldiers adopted facets of the religion of the army and simultaneously held on to their own religious traditions. Material from Dura amply demonstrates that these two aspects of religious practice coexisted peacefully."
278 Peppard 2016, 21–22, 26.
279 Kraeling 1956, 328; Rosenfeld and Potchebutzky 2009, 197; Rajak 2017, 43.
280 According to Rosenfeld and Potchebutzky (2009), it is possible that the synagogue of Dura Europos was erected for a unit of Jewish soldiers serving in the Roman army.
281 On this trilingual inscription in Middle Persian, Parthian, and Greek, see Rubin 2002.
282 Inscription from *The Kaabah of Zoroaster* (ŠKZ), lines 15 and 21. See James 2004, 253.
283 Rosenfeld and Potchebutzky 2009, 197.

FIGURE 6 Julius Terentius fresco from the Temple of Bel: scene with a *vexillum* in the centre and Palmyrene gods in upper left-hand corner

attention towards the highly militaristic iconographic programme of the decoration of the local synagogue.[284] As Stefanie Weisman writes, "like the pagan temples at Dura, the synagogue could well have been a place of hope and spiritual guidance for some members of the Roman army."[285]

The familiarity of Durene Jews with warfare undoubtedly influenced the designers of the synagogue's panels. They incorporated many aspects of contemporary military life into their representations of biblical stories.[286] Much of the iconography is closely related to the aesthetics of the Roman army. Thus, for instance, Israelite warriors are depicted as Roman soldiers by putting *vexilla* in the hands of the Elders and by dressing figures that may represent Solomon's army or holy executioners in legionary uniforms (figure 5).[287] The iconography that appears in the depiction of Julius Terentius from the Temple of Bel, showing a *vexillum*, is very similar (figure 6).[288]

---

284 Rosenfeld and Potchebutzky 2009, 207–217; Weisman 2012, 6; Rajak 2013, 105: "The Dura synagogue paintings are inclined to militarization." On the distinctive traits of Jewish epigraphy related to the military sphere, see Rosenfeld and Potchebutzky 2009, 217–220.

285 Weisman 2012, 31.

286 Weisman 2012, 12–23.

287 Weisman 2012, 31.

288 See Dirven 2007.

FIGURE 7 Dura-Europos Synagogue. The Battle of Eben-Ezer, Scene 2

In the Battle of Eben-Ezer panel, the swords used by the warriors can be identified as *spathae* of the same type as those unearthed in excavations at Dura.[289] In fact, the *spatha*, a long two-edged slashing sword, was the preferred weapon of Roman soldiers on the eastern frontier at the time, as opposed to the shorter *gladius*, which was primarily a stabbing weapon (figure 7).[290] As Stefanie Weisman has noted, "the synagogue artists had an understanding of the correct way to handle this sword: soldiers at the bottom and top of the scene are shown raising their long swords high in the air, in preparation for a slashing motion that will be the death blow for their opponents."[291]

Ted Kaiser believed that the extraordinary decoration of the synagogue illustrating the Hebrew Scriptures in sharp contrast to the interpretation of the Ten Commandments reveals that Jewish communities on the periphery of the Roman world, far away from their homeland, could express and interpret their ancestral religious traditions with a higher degree of freedom, unfettered by notions of orthodoxy.[292] It was these Jews, who were fully integrated in city life and from more "liberal" backgrounds like those indicated by painted synagogue, who proved to be much more inclined to serve as Roman *milites*, without needing to envisage any separate Jewish units.[293] It seems obvious that their

289 Weisman 2012, 8.
290 James, 2004, 140.
291 Weisman 2012, 8.
292 Kaiser 2017, 95.
293 Rosenfeld and Potchebutzky 2009, 207; James 2019, 310–311.

religious scruples must have mellowed considerably. Taking the special features of this Jewish community into account, Stephanie Weisman asked herself the following: "could Jews have served in the Roman army without sacrificing their beliefs and losing their Jewish identity?" According to the evidence, "in the case of Dura-Europos in the third century, the answer seems to be yes."[294]

The triumphal spirit that emerges from the frescoes in the synagogue of Dura-Europos, based on victorious episodes in the Hebrew Bible, seems to reveal a certain competitive spirit with regards to the other religious expressions in which the sphere of the divine was also present. These paintings might reveal a subtle message on the approval of the Hebrew God and the favour he showed to the armies in which Jews served faithfully, following His designs. As it emerges from the victorious Biblical scenes depicted in the synagogue paintings, Jews benefitted from the providential strength of their God, which far surpassed that of other deities.[295]

5 The Presence of Jews in the Imperial Army: Conditions and Historical Evolution

Direct accounts of Jewish individuals who unambiguously displayed their cultural identity within the Roman army are scarce, but by no means nonexistent.[296] Some Jewish soldiers, having adapted well to Greco-Roman culture, did not explicitly recorded their religious or cultural identity, and can only be detected through indirect clues. This suggests that there must have been many other hidden cases that are yet to be revealed by researchers.[297] However, it seems unlikely that those Jews who chose to pursue a career in the military must have necessarily abandoned their religion or concealed their

294 Weisman 2012, 28–29.

295 See Rajak 2013.

296 Some scholars have argued that the limited amount of evidence proves that observant Jews who served in the Roman army were a minority (Castritius 2003b, 3–4; Oppenheimer 2005, 190).

297 As Andrew J. Schoenfeld (2006, 121) rightly pointed out, "Unfortunately, this high level of integration puts the scholar at a certain disadvantage when attempting to identify Roman Jewish soldiers in the historical record. Unless their religion is specifically identified, Jews with Greek or Latin names escape our notice, and it is probable that more Roman Jewish soldiers will be lost to posterity than can ever be identified." For instance, focusing our attention on the epigraphy discovered in Intercisa, Péter Kovács (2010, 175) writes that "obviously people of Jewish origin might have been among the persons with Semitic *cognomina* from Intercisa, but if there is no clear allusion to such a Jewish origin it cannot be proven."

ethnic and cultural background, as some researchers have assumed, inappropriately terming them "renegades".[298] It is true that the Midrash *Genesis Rabbah* mentions a Roman officer under the reign of Hadrian who, being deemed to be an "apostate" or "renegade", seems to have been a former follower of R. Joshua. According to the Midrash, the Roman officer rebuked two of R. Joshua's disciples for their failure to correctly answer three questions on certain biblical texts, telling them that "Your teacher Joshua did not expound it thus."[299] Notwithstanding the difficulties inherent to rabbinical texts as historical sources, there is no evidence of the alleged obligation for Jews to renounce their religion when they joined the army. Being Roman soldiers did not necessarily make them Jewish "renegades".

Although it is likely that some Eastern Jewish communities could have invoked legal precedents to continue to benefit from the exemption from military service, and other Jews had the option to avoid the draft through a substitute (*vicarius*),[300] there seems to have been no obstacle for Jews to voluntarily enlist in the Roman army,[301] either as legionaries, if they were citizens, or as auxiliaries if, by the end of their service, they wished to obtain citizenship and pass it on to their descendants.[302] As Jonathan R.W. Prag states, "military service doubtless provided not only an important outlet for such elements in society, but also a valuable potential means of integration and incorporation."[303] Nevertheless, limited evidence for Jews in the Roman army reveals that there were a number of obstacles to the above. It is very likely that constant political tension (especially in Judaea, during the 1st century CE) in relations between the Jewish people and the Imperial State did not predispose Jews to participate in "foreign" military forces, or Roman authorities to accept them. Nevertheless, this explanation (while certainly not without merit) is not entirely satisfactory, as Shim'on Applebaum has pointed out, as neither did the Empire significantly change its policies towards Jews after the tragic insurrections,[304]

298 Kasher 1985, 80; Applebaum 1989a, 162.

299 *Midrash Bereshit (Genesis) Rabbah*, LXXXII, 8 (Freedman and Simon 1983, II, 757–758). See Rocca 2010, 27; Weisman 2012, 25–26.

300 Pliny the Younger, *Epist.*, X, 30, 1.

301 Cf. Roth 2007, 418; Southern 2007, 132; Olshanetsky 2018, 14; Roux 2020, 27.

302 Schoenfeld 2006, 118. Military diplomas prove that soldiers received the *civitas Romana* after 25 years of service (*qui quina et vicena stipendia meruerant*). See, for instance, Eck 2010. Cf. Lavan 2019a and 2019b. Judging by the diplomas featuring Jewish soldiers Nuno Simões Rodrigues (2007, 134) was wrong to believe that there were no attested cases of this type.

303 Prag 2010, 22.

304 Gaudemet 1988, 346 and 350 ff.; Ayaso, 1990, 98–99; Feldman 1993, 99.

nor had Jews chosen not to serve in the armies of Asia, Egypt or Cyrenaica after Antiochus Epiphanes' aggression against Judaism, much as Caesar's positive attitude made short work of the tension that had arisen from Pompey and Crassus' aggressive behaviour.[305]

There must therefore be other, much deeper reasons to explain Jewish reluctance to serve more actively in the Empire's military structure. In this regard, we should keep in mind that the Roman army was certainly not as "cosmopolitan" as that of the Ptolemies or Seleucids, insofar as the Roman military had a certain atmosphere of "uniformity" around a series of religious rites and practices which the Jews deemed to be idolatrous, such as emperor worship or a number of official ceremonies: *auspicia, signa, auguria*.[306] It is also true that, for the most part, Jewish "leaders" did not find the military to be the best or more natural means for privileged integration in Roman society. Whereas, for instance, the aristocracies of Gaul and Greece joined the officer corps of the Roman army to gain the rank of equestrians[307] and, ultimately, to join the senatorial *ordo*,[308] Jewish elites—except for the Herodians and Tiberius Julius Alexander—never considered this possibility as a direct way to consolidate or further their already privileged status within the Roman Empire. This is particularly surprising considering their former inclination to fight as mercenaries in Hellenistic armies and their obvious proclivity to join the regular armies of the Hasmonean and Herodian dynasties. It is therefore easily understandable that, unlike what happened with the aristocracies of Gaul or Greece, Jews who held well-ensconced positions and were the dominant political group in Judaea—*Syria Palaestina* after the second Jewish War, 132–135 CE—, were never admitted into the ruling class of the Empire with a handful of exceptions, such as the Herodians or the family of Philo of Alexandria.[309]

Thus, it could be said that the main reason for which Jews did not massively join the Roman army was religious, rather than political. Nevertheless, Shim'on Applebaum believes that the difficulties did not stem from the con-

305 Applebaum, 1971, 181.

306 Applebaum, 1971, 181; *Idem*, 1974, 459.

307 However, "the widening of access to equestrian rank within the Roman army contributed to the devaluation of this status over the course of the third century AD." See Davenport 2012.

308 See Tacitus, *Annals*, IV, 4. According to Sara Elise Phang (2012, 78), "In provincial societies a somewhat more prosperous social stratum volunteered or was conscripted for legionary service than in Roman Italy, providing a rationale for provincial recruitment. The local middle classes and even Romanized indigenous aristocrats thus sought affiliation with the imperial power through military service." Cf. Le Bohec 2000, 76–77, 89–93, 97–99.

309 See Goodman 1987, 48–49.

ditions imposed by the practices of Judaism as much as from the rejection of the religious rites that determined and justified Roman military life and organisation.[310] It is very likely that, as Martin Goodman points out, this reality would have changed if Jews had been allowed to serve more widely in ethnically-separate units (*contubernia*), as had presumably happened during the Hellenistic period. Obstacles to obtaining kosher food or drinking wine in the company of gentiles would have been reduced considerably[311] and, should it have been strictly necessary, a specific unit remaining inactive on Sabbath would not have been an insurmountable problem. Likewise, pagan rituals and symbolic formalities of a military nature could be avoided relatively easily by those who were not in the officer corps, which was much more subject to ceremonies and formalities, as Jews were likewise exempted from the imperial cult.[312]

We cannot rule out that these circumstances may have been met in very specific cases. Nevertheless, it seems logical to assume that, as in many other fields, the immense majority of Jews who became fully integrated in the Roman army displayed a high degree of cultural assimilation and an inevitable slackening of some of the ritual practices of their religion. In this case, their cultural isolation within a larger and more diverse social structure would end up blurring their own heritage, even if they did not renounce it, and that is probably why it is sometimes extremely difficult to detect the presence of Jewish identity in clearly gentile contexts.[313] Tiberius Julius Alexander, a nephew of Philo of Alexandria who had a brilliant career in the Roman army, provides us with an example of this. Even though there are no direct accounts certifying his previous military service or *tres militiae* (*praefectus cohortis*, *praefectus alae* and *tribunus angusticlavius*), which was usually required from those who wished to pursue an equestrian career, his military experience is implied by his active participation in highly relevant operations, first alongside Gnaeus Domitius Corbulo (whom Nero appointed chief of staff in 63 CE)[314] in his campaign against the Parthians in Armenia and, later on, with Titus in Jerusalem. This is one of the few known cases of a member of the Jewish aristocracy choosing a career in the military as the most effective means of rising to the highest spheres of Roman society. We are unaware of whether or not this decision required him

310 Applebaum 1974, 459.
311 Goodman 1987, 48 note 32. Cf. Wilde 1949, 30; Rosenblum 2016, 71.
312 Hild 1884–1885, 29; Askowith 1915, 202; Applebaum 1974, 458.
313 Barclay 1996a, 325. Cf. in general Rutgers, 1998a, 39–40.
314 Syme 1970.

to renounce his religion. The fact is that, as we mentioned above, Flavius Josephus states that he lacked the religious piety of his father, as "he did not stand by the practices of his people" (τοῖς γὰρ πατρίοις οὐκ ἐνέμεινεν οὗτος ἔθεσιν).[315] In any event, it is possible that his ancient beliefs and customs, such as the Sabbath rest or dietary laws, would have been a serious obstacle to his ambitious plans, and that he felt it was best to adapt to the ritual forms of paganism.[316] His position as *epistrategus* of the Thebaid in 41 CE, as procurator of Judaea (46–48 CE) and prefect of Egypt from 66 CE onward clearly required him to carry out a number of official duties of a religious nature that were hardly compatible with Judaism.[317]

According to John M.G. Barclay,[318] the attitude adopted by Antiochus, the son of the main Jewish magistrate (ἄρχων) of Antioch,[319] at the outbreak of the Jewish revolt, reveals a similar case. When Vespasian arrived to take command of the army stationed in Judaea (67 CE), Antiochus accused a number of Jews of conspiring to burn the city during the night. Josephus' narrative makes more sense if we assume that this figure who, being under the governor's protection, had been educated by soldiers, held the rank of officer in the military.[320] Only thus could we properly explain the authorities' willingness to give him command of Roman troops.[321] We can assume that a Jew who held a high position in the military would have been faced with a dilemma when it came to preparing the fight against his brethren. It is therefore obvious that Antiochus must have somehow renounced his ancient traditions when he decided to accuse a group of Jews of being "revolutionaries". In fact, he proved his "conversion to paganism" (μεταβολὴ) and his hatred of Jewish customs by offering sacrifices "in the manner of the Greeks",[322] and attempting to convince other Jews to do the same.[323] In fact, this renunciation of Judaism is the most extreme case of "cultural assimilation" known to us.[324]

315 Josephus, *Ant. Iud.*, XX, 100.
316 Kasher 1985, 78.
317 Étienne 2000, 130–131.
318 Barclay 1996, 256.
319 Josephus (*Bell. Iud.*, VII, 47) states that he was very respected among the Jews on account of his father's good reputation.
320 Without considering that he may have been a military man, Edith Mary Smallwood (1976, 361–362) believed that Antiochus held Roman citizenship and was even a magistrate.
321 Josephus, *Bell. Iud.*, VII, 52.
322 Josephus, *Bell. Iud.*, VII, 50: [...] τὸ ἐπιθύειν ὥσπερ νόμος ἐστὶ τοῖς Ἕλλησιν.
323 He also banned Jewish observance of the Sabbath (Josephus, *Bell. Iud.*, VII, 51–53).
324 Barclay, 1996, 322–323. At the time when the revolt broke out, Josephus writes about the ἀπόστασις of the Jews in Scythopolis, who chose to fight alongside the local Greeks in their confrontation with Jewish enemies from outside (*Bell. Iud.*, II, 466–476; *Vit.*, 26), although

Aside from the difficulty inherent to discovering Jewish identity when it has been drastically hidden or omitted in sources available to us, there is also information on the presence of Jewish soldiers in the Roman army which, despite its vague or accidental nature, is interesting to us as an eyewitness account. An example is the speech addressed by emperor Marcus Aurelius to his western troops in 175 CE to steel their hearts for the looming clash with Avidius Cassius and his army of easterners, among whom were Jews. It is even likely that, in this case, the Jews had the opportunity to serve as *auxilia* in specific ethnic units.[325] In any event, judging by the tone of the speech, Jewish soldiers serving in the Roman army among other "foreigners" are mentioned as an ordinary phenomenon:[326]

> But since Cassius would never consent to adopt this course,—for how could he trust me after having shown himself so untrustworthy toward me?—you, at least, fellow-soldiers, ought to be of good cheer. For surely Cilicians, Syrians, Jews, and Egyptians have never proved superior to you and never will, even if they should muster as many tens of thousands more than you as they now muster fewer […].[327]

Even if we were to assume that this speech was reworked by Cassius Dio according to the image of the Roman army made up from Eastern troops, as it was at the time he wrote (ca. 230 CE), and transposed to an earlier period, nothing pre-

---

this decision seems to have been political rather than religious in nature (Barclay 1996, 323 note 5). The notion of "apostasy" is still highly ambiguous at this period in time. Excluding those cases in which it applied to those who expressly and formally renounced Judaism—as seems to have been the case of this Antiochus—, it is difficult to identify the boundaries established within a given community on the degree of loyalty to religious traditions that was tolerable or required. According to that author, "Jews could and did differ in where they considered other Jews to have 'crossed the boundary' out of Judaism. To observe 'apostasy' in the Diaspora is to observe how Jews in the Diaspora considered other Diaspora Jews to have stepped outside (what they took to be) legitimate Judaism" (Barclay 1998, 81). For instance, in *3 Maccabees*, a late-1st century CE Alexandrian work, acquiring Greek citizenship was tantamount to "apostasy", even though this belief was not shared by all Diaspora Jews. See Barclay 1998, 83–84; Collins 2000, 122–131.

325 Rocca 2010, 26.

326 Oppenheimer 2005, 190. Cf. Juster 1914, II, 273; Olshanetsky 2018, 16.

327 Cassius Dio, *Epitome of book* LXXII (LXXI), 25, 1: Ἐπεὶ δ' οὐκ ἄν ποτε συγκαθεῖναι ἐς τοῦτο ὁ Κάσσιος ἐθελήσειε (πῶς γὰρ ἂν πιστεύσειέ μοι, ἄπιστος οὕτω περὶ ἐμὲ γεγενημένος), ὑμᾶς γε ὦ συστρατιῶται χρὴ θαρρεῖν. Οὐ γάρ που κρείττους Κίλικες καὶ Σύροι καὶ Ἰουδαῖοι καὶ Αἰγύπτιοι ὑμῶν οὔτε ἐγένοντό ποτε οὔτε ἔσονται, οὐδ' ἂν μυριάκις πλείους ὑμῶν, ὅσῳ νῦν ἐλάττους εἰσίν, ἀθροισθῶσιν. […]

cludes us from using his account as yet more evidence of the presence of Jewish soldiers serving in the Roman army.[328]

Twenty years later, Jews of Palaestina fought in the army of Pescenius Niger against Septimius Severus. According to Aelius Spartianus, the alleged author of the *Vita Severi*, this emperor "revoked the punishment which had been imposed upon the people of Palestine on Niger's account" a few years after he achieved power.[329]

Some late sources record a Jewish and Samaritan war or revolt which allegedly took place after Niger's final defeat in the year 194.[330] Nevertheless, it is surprising that none of the most relevant contemporary sources, such as Cassius Dio and Herodian, or any of the later ones that tended to include such episodes in their narratives, such as the *Historia Augusta* itself, mention such a remarkable fact. The fact that rabbinical literature, which could be expected to comment on the subject, is also silent on the matter, casts doubt on its historicity.[331] According to Michael Avi-Yonah,[332] this historiographical tradition could be a reflection of distant memories of an alleged local conflict between the Samaritan followers of Niger and Severus' Jewish supporters in Galilee. Nevertheless, the apparent division of alliances surrounding the two contenders that emerged in the cities of Samaria would disprove this hypothesis.[333] In fact, it could be said that an immense majority of Jews and Samaritans wholeheartedly supported Septimius Severus.[334]

It is truly difficult to assess the significance of the presence of Jews within the Roman army, either in specific military units or on an individual basis.[335] It would be unwise to unhesitatingly accept the notion that, as Jean Juster argued, Jewish military service in the Empire had become commonplace.[336] It is worth keeping in mind that the mere acknowledgement of a reality is not grounds in itself for assigning it a greater level of importance than that which it deserves in

328 Oppenheimer 2005, 190.

329 *Scriptores Historiae Augustae*: *Sev.*, 14, 6: [...] *Palaestinis poenam remisit quam ob causam Nigri meruerant* [...]. See Juster 1914, II, 273 note 3.

330 Orosius, VII, 17, 3; Eusebius of Caesarea, *Chron.*, *olympiad* 244 (*ad annum* 197), 5d, ll. 19–20: *Iudaicum et Samariticum bellum motum*. See Smallwood 1976, 488–489.

331 According to Martin Goodman (2007, 498–499 = 2008, 475–476), it is likely that this report of an alleged Jewish-Samaritan war reflects Jews' reputation as habitual rebels against the Empire, especially in Christian circles.

332 Avi-Yonah 1984, 77–78.

333 Smallwood 1976, 489.

334 González Salinero 2019a, 238.

335 Rocca 2010, 27; Olshanetsky 2018, 22.

336 Juster 1914, II, 265–279.

all fairness. Nevertheless, accounts on the subject do not allow for a reduction of the Jewish presence in the Roman army to a merely marginal phenomenon, nor can its minority nature be dismissed as an exceptional and insignificant episode.

That being said, regardless of our inclination concerning the importance to be accorded to the Jewish element within the Roman military, we cannot deny the evidence that results from ancient sources. It is therefore all the more interesting to consider the true reasons that gave rise to this phenomenon. We have already mentioned the apparent incompatibility between correct observance of Jewish religious practices and the requirements of Roman military life, and ways in which a circumstantial solution may have been found.[337] We must likewise keep in mind that the "flexibility" of the Roman State would have allowed Jews who were unwilling to renounce the ritual practices of their religion (who were certainly a minority) to integrate fully in the army, notwithstanding the issues such a "forced" situation might entail.[338] In theory, certain pagan ceremonies and rites such as sacrifices, the oath (*sacramentum militiae*) and the imperial cult were significant hurdles to Jewish religious conscientiousness.[339] In practice, however, we come across cases of Jews who did not hesitate to swear on the emperor's name: a papyrus dated 10 February 101 CE features a declaration of death in which a Jew named Soteles, son of Josepos, swore on the emperor's name,[340] and this formula appears once again with the added term τύχη (equivalent to *genius*), in another papyrus dated 127 CE.[341] Nevertheless, the military oath, as it appears described by 4th-century Roman grammarian

337 See also Appendix 1.

338 Shean 2010, 182. Cf. Gichon 2009, 150–151. In contrast, Oppenheimer 2005a.

339 See Applebaum 1971, 181; Kasher 1985, 78; Smallwood 1976, 137 note 62; Nemo-Pekelman 2010, 177–178. On the religious aspects of the Roman army, see, as an introduction, the following studies: Nock 1952; Richmond 1962; Helgeland 1978, 1495 ff. Even though the case has often been made for the significance of the imperial cult as a unifying element in the Roman army, the oath, which was not only commonplace in the social and legal organization of Rome, but also in its army (*sacramentum militiae*), was no less important: see in general Rüpke 1990, 76–80; Zuccotti 2000, 21 ff. and 83; Letta 2020, 56–57.

340 *CPJ II*, 427, ll. 18–24 = *BGU IV*, 1068: Σωτέλης Ἰω|σήπου ὁ πρωγε|γραμένος ὠμ|νύω Αὐτοκράτο|ρα Καίσα[ρα Νέρουαν | Τραιαν[ὸν | Σεβα[στόν ("Soteles, son of Josepos, the one mentioned before, I swear by the Emperor Caesar Nerva Trajan Augustus"). See Cotton 1998, 168.

341 *P.Yadin* 16, ll. 33–35 (Lewis, Yadin and Greenfield 1989, 67): Βαβ|θα Σίμωνος ὄμνυμι τύχην κυρίου Καίσαρος καλῇ πίστει ἀπογε|γράφθαι ὡς προγέγραπ[τα]ι. ("I, Babtha daughter of Simon, swear by the *genius* of our lord Caesar that I have in good faith registered as has been written above".) See Lewis 1991; Varon 2001, 272 note 8; Kraemer 2004, 147–148 (No. 62D).

Servius Honoratus, may not necessarily have been an insurmountable obstacle for Jews, as new recruits simply had to swear that "they would act on behalf of the Republic, and would not leave until they had served their time".[342] It is even likely that the exemption from taking the solemn oath of loyalty and from emperor-worship that was granted to Jews and mentioned by Flavius Josephus in *Contra Apionem* was extended through time:

> He [Apion] should rather have admired the magnanimity and moderation of the Romans in not requiring their subjects to violate their national laws, and being content to accept such honours as the religious and legal obligations of the donors permit them to pay. They are not grateful for honours conferred under compulsion and constraint.[343]

Nevertheless, it was deemed acceptable for sacrifices related to the imperial cult to be substituted by sacrifices to Yahweh for the health of the emperors and the people of Rome.[344] Josephus described it as follows:

> He did not, however, forbid the payment of homage of another sort, secondary to that paid to God, to worthy men; such honours we do confer upon the emperors and the people of Rome. For them we offer perpetual sacrifices; and not only do we perform these ceremonies daily, at the expense of the whole Jewish community, but, while we offer no other victims in our corporate capacity, even for the [imperial] family, we jointly accord to the emperors alone this signal honour which we pay to no other individual.[345]

---

342 Servius Honoratus, *Comm. ad Aen.*, 8, 1: [...] *legitima erat militia eorum, qui singuli iurabant pro republica se esse facturos, nec discedebant nisi completis stipendiis, id est militiae temporibus: et sacramentum vocabatur, ut "fas mihi Graiorum sacrata resolvere iura".* [...]. See Watson 1985, 49.

343 Josephus, *Contr. Ap.*, II, 73: [...] *cum potius debuerit ammirari magnanimitatem mediocritatemque Romanorum, quoniam subiectos non cogunt patria iura transcendere, sed suscipiunt honores sicut dare offerentes pium atque legitimum est. Non enim honoribus gratiam habent qui ex necessitate et uiolentia conferuntur.* See Hild 1884–1885, 29; Smallwood 1976, 137 and 378–381. In fact, there were no laws forcing Jews to participate in pagan rituals, nor are there any accounts of Jews being compelled to do so (Olshanetsky 2018, 24). Cf. Czajkowski 2015.

344 Sartre 1991, 394.

345 Josephus, *Contr. Ap.*, II, 76–77: *Aliis autem honoribus post deum colendos non prohibuit uiros bonos, quibus nos et imperatores et populum Romanorum dignitatibus ampliamus. Facimus autem pro eis continua sacrificia et non solum cotidianis diebus ex impensa communi omnium Iudaeorum talia celebramus, uerum cum nullas alias hostias ex communi*

ΥΠЄΡ CωΤΗΡΙΑC ΤωΝ Κ[. . .]

ωΝ ΗΜωΝ ΑΥΤΟΚΡΑΤωΡω[.]

ΚΑΙCΑΡωΝ Λ · CЄΠΤ · CЄΟΥΗ[. . .]

ЄΥCЄΒ · ΠЄΡΤ · CЄΒ · ΚΑΙ Μ · ΑΥΡ · Α[. . . . .]

ΙΝΟΥ [. . . . .] ЄΤΑ ΥΙωΝ ΑΥ[. . . . .]

ЄΥΧΗC ΙΟΥΔΑΙωΝ

FIGURE 8 Greek inscription from the synagogue at Kasyon (Kitzion), near Safed, Galilee (circa 197 CE). *CIJ II*, 158 (n.º 972).

In this regard, a valuable account of support and loyalty towards imperial power can be found in a Greek inscription which was discovered in the ruins of the synagogue at Kasyoun or Qeisoun (Kitzion or Qasioun), near Safed, in Galilee (figure 8). In this inscription, some Jews paid homage to the emperor and his family (his wife Julia Domna and his children, Caracalla and Geta) circa 197 CE through the formula ὑπὲρ σωτηρίας τῶν κυρίων ἡμῶν αὐτοκρατόρων καισάρων, which is equivalent to the Latin expression *pro salute Augusti domini* ... found in the Cosmius inscription:[346]

> For the welfare of our lords, the emperors and caesars Lucius Septimius Severus (the empress Julia Domna), Marcus Aurelius Antoninus ("Caracalla") and Lucius Septimius Geta, his sons, a dedication ex voto of the Jews.[347]

We could assume that they may have enjoyed a number of "privileges" that would have allowed them to abstain from, or at least subtly avoid other "idolatrous" practices. In fact, the existence of certain peculiarities, such as the right granted to Mithraists to carry triumphal crowns in their hands rather than wear them on their heads (an honour they only paid to their god) is a clear example of the efforts made by Romans to reconcile various religious beliefs

neque pro filiis peragamus, solis imperatoribus hunc honorem praecipuum pariter exhibemus, quem hominum nulli persoluimus. Cf. Josephus, *Bell. Iud.*, II, 197.

346 *IJO I*, Pan3 (*cf. CIL* III.1, 3327; *CIL* III.Suppl. 10301; Erdélyi and Fülep 1954, 323, No. 329; *CIJ I*, 677; *RIU* V, 1991, 1051). Cf. *IJO II*, 124. See Friedheim, 2021, 296.

347 *CIJ II*, 972: Ὑπὲρ σωτηρίας τῶν κ[υρί]ων ἡμῶν αὐτοκρατόρω[ν] καισάρων Λ. Σεπτ. Σεουή[ρου] Εὐσεβ. Περτ. σεβ. καὶ Μ. Αὐρ. Ἀ[ντωνε]ίνου [καὶ Λ. Σεπτ. Γ]έτα, υἱῶν αὐ[τοῦ ἐξ] εὐχῆς Ἰουδαίων—καὶ Ἰουλίας Δόμνης Σεβ[αστῆς]. See Smallwood, 1976, 496–497; Avi-Yonah 1984, 77; Grant, 1999, 269; *IJO II*, 514–515; De Bonfils 2005, 54–55; Hadas-Lebel 2009, 195; González Salinero 2019a, 238–239; Friedheim, 2021, 293–299.

and military service.[348] It is obvious that, in this case, the fact that Commodus was the first emperor to be initiated in the mysteries of Mithras and took the Mithraic titles of PIVS FELIX INVICTVS[349] had something to do with it, but other emperors also displayed an obvious inclination towards Judaism. Under his successors in the Severan dynasty, Mithraism held a pre-eminent position within the Roman religious world, though the Jews achieved a suitable and profitable *modus vivendi* under them.[350] It could be stated that further efforts were made to facilitate Jewish military service during this period (early 3rd century CE), considering that the removal of any hurdles that precluded them from accessing local administrations was likewise accompanied by the formulation of a flexible policy that would foster an increased compatibility between Judaism and military life.[351] To this end, and as it was stated in the new regulations, it was necessary that the formalities required to hold public office would be null and void whenever they clashed with the tenets of Judaism. The *Digest* provides us with a description of the law penned by Ulpian:[352]

> The same author, On the Proconsul's office, Book Three ... The Divine Severus and Antoninus permitted those that follow the Jewish religion to enter offices, but also imposed upon them liturgies such as should not transgress their religion.[353]

It seems that, from the moment in which Roman citizenship was granted to all free men of the Empire through the passing of the *Constitutio Antoniniana* of 212 CE, there was a significant increase in the military service of Jews in the Roman army.[354] Even though sources available to us are quiet on the subject, there are a number of clues that point in this direction. It is remarkable that pagan historian Zosimus (second half of the 5th century) records that Palestinians were perfectly integrated in the Roman army of the mid-3rd cen-

348 Tertullian, *De corona*, XV, 3–4. See Demougeot 1961, 358 ff.; Helgeland 1978, 1498.

349 Demougeot 1961, 359 and 363.

350 Applebaum 1971, 182; González Salinero 2019a.

351 Applebaum 1971, 182.

352 One of the more important jurists in the early third century, who held several important government offices in 222–223 CE.

353 *Ulpianus, Digest* 50.2.3.3: [...] *Iter libro tertio de officio proconsulis ... Eis, qui Iudaicam superstitionem sequuntur, divi Severus et Antoninus honores adipisci permiserunt, sed et necessitates eis imposuerunt, qui superstitionem eorum non laederent.* See Gichon 2009, 154; Weisman 2012, 25 and 31.

354 See Scheiber 1983, 725–728; *IJO III*, 34; Schoelfeld 2006, 122; Olshanetsky 2018, 11.

tury,[355] although it is uncertain whether those soldiers were Jewish, Christian or both.[356] At any rate, according to the *Historia Augusta*, there were enough Jews in the Roman army at the time to participate in the dedication of a multilingual inscription—which featured the Jewish language, namely Aramaic or Hebrew—on a tomb erected for Gordian III by his soldiers:

> The soldiers built Gordian a tomb near the camp at Circesium, which is in the territory of Persia, and added an inscription to the following effect in Greek, Latin, Persian, Hebrew, and Egyptian letters, so that all might read: "To the deified Gordian, conqueror of the Persians, conqueror of the Goths, conqueror of the Sarmatians, queller of mutinies at Rome, conqueror of the Germans, but no conqueror of Philippi."[357]

On the other hand, it is very significant that, according to the *Res Gestae Divi Saporis*, an inscription found in Persia, at Naqsh-e Rustam, in the army that the Emperor Valerian used some years later in his expedition against the Persians there were troops coming from Judaea.[358] It is true that the mention in the inscription of all these "nations" (Germany, Raetia, Norica, Dacia, Iberia, Africa, Moesia, Phoenicia, Thrace, Bithynia, Asia, Pamphylia, Isauria, Lyconia, Galatia, Lycia, Cilicia, Cappadocia, Phrygia, Syria, Arabia, Mauretania, etc.) could fulfill a purely propagandistic function whose objective would not be precisely to incorporate into the text an exact description of the diversity of the ethnic components of the Roman army, but rather the praising the triumph over a power that was capable of bringing together so many peoples, but that was insufficient to defeat the all-powerful army of the great Sapor I. Therefore, it could be thought that it was a mere rhetorical resource far from reality. However, we know that each and every one of these peoples mentioned in the *Res Gestae Divi Saporis* served, in fact, at that time in the Roman army, while it is strange that others do not appear that, such as the Egyptians, the Ituraeans or the Tyrians, could reinforce that supposed purely propaganda

355 Zosimus, *Nov. Hist.*, I, 52, 4 and 53, 2.

356 Varon 2001, 276.

357 *Scriptores Historiae Augustae*: *Gordiani tres*, 34, 2–4: *Gordiano sepulchrum milites apud Circesium castrum fecerunt in finibus Persidis, titulum huius modi addentes et Graecis et Latinis et Persicis et Iudaicis et Aegyptiacis litteris, ut ab omnibus legeretur: "Divo Gordiano, victori Persarum, victori Gothorum, victori Sarmatarum, depulsori Romanarum seditionum, victori Germanorum, sed non victori Philipporum."* Cf. Stern, 1976–1984, II, 634 (No. 525). See Rocca 2010, a28; Olshanetsky 2018, 16. Ammianus Marcellinus (XXIII, 5, 7) also remembers that Julian and his army saw the tomb (or cenotaph) of Gordian.

358 *RGDS* III, 23 (Maricq 1958, 311). Cf. Rostovtzeff 1943, 28–30.

purpose. Thus, everything would indicate that those peoples who, unlike the others, were absent from the list, did not actually participate in that military campaign and, therefore, could not be verified by the Persians, which would imply that the information regarding the nations mentioned in the entry is reasonably true.

In short, epigraphic sources attributable to that period and to the military sphere, in which the synagogue of Dura Europos was fully integrated, must lead us to acknowledge the increasing presence of Jews serving in the Roman army.

CHAPTER 4

During the Later Roman Empire

1 Material Evidence

1.1 *The Levant*

A fourth-century CE epitaph of the grandson of a Jewish officer in the Roman army featuring a Greek and Hebrew inscription, which was found in the Ioppe necropolis (Jaffa, Israel), was displayed in an exhibition in the Israel Museum in 1972 and subsequently lost (figure 9). Fortunately, a number of photographs that allow us to easily read the epigraphic text are available to us:

> ΘΑΝΟΥΜΥΙΟΣ
> ΣΙΜΩΝΟΣΕΝ
> ΓΟΝΙΝΒΕΝΙΑ
> ΜΙΝΤΟΥΚΕΝ
> ΤΗΝΑΡΙΟΥΤΗΣ
> ΠΑΡΕΝΜΒΟΛΗΣ
> שלום (lulav)

> Thanum son of Simon, grandson of Beniamin the centenarius from Parembole ('camp'). Peace (lulav).[1]

The inscription is unambiguous about the fact that Benjamin, Thanum's grandfather, was a soldier. The fact that it states his rank (κεντηνάριος, *centenarius*) points to the military's high social standing at the time.[2] Indeed, as Walter Ameling notes, "the grandfather Benjamin is probably named in the epitaph because of his distinction as centenarius."[3] We do not know the exact location of the camp in which he served. William Horbury and David Noy believed that "this is probably the one just south of Syene on the upper Nile, mentioned as a *statio* in the Antonine Itinerary 161W."[4]

---

1 *CIIP III*, No. 2240 (*cf. CIJ II*, No. 920; *JIGRE*, No. 147): Θανουμ υἱὸς | Σίμωνος ἐν|γόνιν Βενια|μὶν τοῦ κεν|τηναρίου τῆς | Παρε{ν}μβολῆς | שלום (shalom). See Applebaum 1971, 182; Varon 2001, 276; Oppenheimer 2005b, Rocca 2010, 29 note 48.

2 On the military rank of *centenarius*, see Christol 2006.

3 *CIIP III*, 119.

4 *JIGRE*, 239–240. Barrington's Atlas (Talbert 2000) lists three places by this name in Egypt (73 D4, 75 E2, 81 C2), one in Turkey (87 F4) and another in Macedonia (49 D2).

 | DOI:10.1163/9789004507258_006

FIGURE 9
Greek and Hebrew inscription of Thanum from the Ioppe necropolis (Jaffa, Israel). CIIP III, 119 (n.º 2240).

FIGURE 10
Greek and Hebrew inscription of Eusebius from the Ioppe necropolis (Jaffa, Israel). CIIP III, 42 (n.º 2177).

The same necropolis in Jaffa has yielded the recently-edited inscription of the *magistrianus* and *presbeutés* Eusebius (figure 10), which has been dated to the same period (4th–5th centuries):

ΕΥΣΕΒΙΟΥ
ΜΑΓΙΣΤΡΙ
ΑΝΟΥΚΑΙ
ΠΡΕΖΒΕΥ
ΤΟΥ שלום (lulav)

(Tomb) of Eusebius, magistrianus and presbeutes. Peace (lulav).[5]

5 *CIIP III*, No. 2177: Εὐσεβίου | μαγιστρι|ανοῦ καὶ | πρεζβευ|τοῦ. שלום. Other *magistriani* are recorded in Iudaea/Palaestina: in Bet Yeraḥ (Philoteria), Ḥammath Gader and Gerasa. See *CIIP III*, 43.

FIGURE 11
Depiction of *agentes in rebus* or *magistriani*. Mosaic from the Casale di Piazza Armerina (Sicily). *Grande Caccia*, 36b, mosaic III B
SOURCE: CARANDINI, RICCI AND DE VOS 1982, FIG. 18

Eusebius held two titles, one of them being μαγιστρίανος, which refers to a post in the Roman administration, and the other being πρεσβευτής, probably a position within the Jewish community. As Walter Ameling states, "*magistrianus* was an administrative position counted among the *agentes in rebus*, under the aegis of the *magister officiorum*."[6] *Agentes in rebus*—i.e. "those in charge of matters," also known as *curiosi* or κουρίωσοι—were an elite military force in charge of watching over State security in the provinces of the Empire (figure 11).[7] Even though they were assigned to serve the civil administration, both the organisation and ranks of the *agentes in rebus* were military in origin.[8] As in the army, there were a number of different ranks and titles: *ducenarii*, *centenarii*, *biarchi*, *circitores* and *equites*. Much like a soldier, an *agens in rebus* was deemed to be a *vir strenuus* or *vir devotus* (in Greek, καθωσιωμένος),[9] though some of them had the title of *clarissimi*.[10]

We cannot rule out a relationship between this inscription and the one from Thanum if we accept the possibility that the *centenarius* Benjamin also

6 *CIIP III*, 43. Cf. *Nov. Iust.*, 17, 4 and the inscription of *magistrianos* Lucius Aurelianus from Sardis (*CIG*, 3467). See Clauss 1980, 24; Delmaire 1995, 98; Di Paola 2005, 67 and 71; González Salinero 2008; Di Paola 2009; *Eadem* 2013, 306–310.

7 It is likely that *agentes in rebus* acted as chiefs of provincial military bureaus from the outset (Le Bohec 2006, 92). In fact, they were entitled to directly request assistance from the military while carrying out their missions (Delmaire 1995, 116).

8 Linder 1987, 223–224 note 4; Rabello 1987–1988, II, 696; Stemberger 2000, 305. Its genuinely military nature is reflected in Libanius, who sometimes calls them soldiers (στρατιώτης). See Herz 2002, 325.

9 Di Paola 2005, 19 and 102–103.

10 According to *CTh.*, 6.27.12, the *agentes in rebus* achieved senatorial rank in 399.

belonged to the *schola agentum in rebus*,[11] as William Horbury and David Noy have suggested.[12]

An inscription from Aphrodisias, dated to the rule of Constantine at the earliest, features a former *primipilaris* using the uncommon formula of the "gifts of God" that are returned to Him:

> To God who listens, Fl(avius) Eusebius, *e primipilaribus*, from the gifts of God made the first and third intercolumniation.[13]

The expression ἀπὸ πριμι|πιλαρίων (equivalent to *e primipilaribus*) indicates that Flavius Eusebius had served in the army as *primipilaris* before retiring.[14] He seems to have adopted the name Flavius under the reign of Constantine at the earliest.[15] The latest reappraisal of this inscription based on a study of the expression Θε[.]ῷ ἐπηκόῳ—which originated in Jewish circles—as it relates to the notion of δόμα or its plural δόματα (for a gift or gifts)—which has Jewish epigraphic parallels in other locations, such as Sardis—, has allowed us to establish that Flavius Eusebius was most likely Jewish.[16] Nevertheless, Joyce Reynolds and Robert Tannenbaum, being unable to conceive the notion of Jews serving in the civil administration or in the imperial army, baselessly assumed "that Eusebius' office was honorary rather than actual."[17]

Moving on to a later period—the 6th century CE—, we find a portrayal of a Roman soldier on a mosaic in the synagogue of Meroth (Northern Galilee) featuring an Aramaic inscription with the name "Yudan son of Shimon, *mny* [one silver *maneh*?]" (figure 12). The warrior is dressed in a white long-sleeved tunic with a red cloak draped over his shoulders and exhibits typically Roman weaponry. According to Rhoda Terry and Steven Fine, "the design on the *insignia* closely resembles the emblem of the Victores, a Roman legion." These unambiguous traits have led these scholars to conclude that "the model for this image was a Roman soldier."[18] We are unaware of the reason why Yudan ben

11 The genitive plural *agentum* was much more common than *agentium* in official terminology (for instance, *CTh*. 6.27.12). See Clauss 1980, 23, note 35.

12 *JIGRE*, 240. *Cf.* *CIIP III*, 120. Cf. Delmaire, Rougé and Richard 2005, 393 note 2.

13 *IJO II*, No. 19 = *SEG* 37, No. 851 = Reynolds and Tannenbaum 1987, No. 9 = Roueché 1989, No. 10: [.]Θε[.]ῷ ἐπηκόῳ Φλ(άβιος) | Εὐσέβιος ἀπὸ πριμι|πιλαρίων ἐκ τῶν | τοῦ θεοῦ δομάτων | τὸ πρῶτον καὶ τρίτον | διάστυλον ἐποίησεν.

14 On the rank of *primipilaris*, see Dobson 1974.

15 Reynolds and Tannenbaum 1987, 137.

16 Rajak 1998a, 236–237; Varon 2001, 276.

17 Reynolds and Tannenbaum 1987, 137.

18 Terry and Fine 1996, 167.

FIGURE 12
Mosaic depiction of a Roman soldier with the Aramaic inscription "Yudan son of Shimon, *mny*." Synagogue of Meroth (Northern Galilee)
PHOTO: STEVEN FINE IN *CENTER FOR ONLINE JUDAIC STUDIES* (WWW.COJS.ORG/ICONOCLASM)

FIGURE 13
Notitia Dignitatum Orientis, XXXVI, 7 (*Syna Iudeorum*)
SOURCE: NEIRA FALEIRO 2005, 273

Shimon was given such a privileged portrayal. The fact that he is identified by his name reveals that he held high status in the community. It may be that he funded the construction of the synagogue as a *patronus*, or that his successful career in the military, which may have provided him with a comfortable lifestyle, benefitted the religious community to which he felt so close.[19]

19 Cf. Rocca 2010, 29.

1.2 *Egypt*

The *Notitia Dignitatum Orientis* (ca. 425–429 CE) contains a brief reference to the encampment of *Cohors prima Epireorum* in *Castra Iudaeorum* ("camp of the Jews"),[20] which was located in the *Augustamnica* district east of the Delta.[21] This is clearly a fortress that owed its name to a former Jewish military colony. Edouard Naville convincingly argued that it was the former colony of Onias, which, according to Josephus, was precisely located in that same district. The archaeological remains of a military settlement in Tel el-Yehoudieh have been identified as *Castra Iudaeorum*.[22] The "village of the Jews" (*Vicus Iudaeorum*) mentioned in *Itinerarium Antonini*, which is somehow related to it, has been located to the north of it.[23] It is likely that this "Jewish camp" was still occupied by Jewish soldiers under the Roman Empire, but we cannot ascertain whether they remained there in active service or if they shared it with other soldiers when *Cohors prima Epireorum* was stationed there. This may not be the only case in which the presence of Jews can be found in the lists of soldiers from military encampments. The Roman fort at Tell el-Herr (*Magdolum*), in the eastern Delta, has yielded twelve *óstraca*, dated to the 4th century CE, which feature soldiers' names alongside accounting entries. These names include Abraam and Moses—*Moyse*, in the text—, both of whom were Jewish members of the auxiliary forces assigned to the garrison.[24]

One of these names appears in a document from Philadelphia (in the Fayyum area) dated c. 342–351:

> To my lord and patron the praepositus, (from) the mother of Moses. You sent for Heron, lord patron, so he went with the barbarians. After God we have no help other than you. So I sent Athioeis to you, my lord. I beg your feet, lord patron, since you know that "the five days are the whole year", I beg you and beseech you, lord, to permit him the few days. If you order something, lord, tell Athioeis. I pray for your health, lord patron.[25]

---

20 *Not. Dig. Orientis*, XXVIII, 42 (Neira Faleiro 2005, 246 = Seeck ^{3}1983, 60). Later in the text we also find the *Equites promoti indigenae et Syna Iudaeorum* in Osrhoene, probably near *ad Tygrem* or close to Lake *Scodra* (?): *Not. Dig. Orientis*, XXXVI, 7 and19 (Neira Faleiro 2005, 273 and 235) (figure 13). See Applebaum 1971, 182; Guichaoua 2019, 202 and 204.

21 Cf. Josephus, *Ant. Iud.*, XIV, 133; *Bell. Iud.*, I, 191; *CPJ I*, 147–178.

22 Naville 1890, 19–20; Kasher 1985, 120–122; Schürer 1987, III, 48–49; Varon 2001, 271 note 1.

23 *Itinerarium Antonini*, 42 (Cuntz 1990, 23).

24 See Valbelle and Carrez-Matraray 2000, 68, 146, 151; Perea Yébenes 2020, 282–283.

25 *P.Lond.*, II, 410 = *P.Abinn.*, 34 (86): τῷ δεσπότῃ μου | καὶ πάτρωνι πραιποσίτῳ | ἡ μήτηρ τοῦ Μουσῆ. | ἔπεμψας ἐπὶ Ἥρωνα, | κύριε πάτρων. ἀπῆλθεν | οὖν μετὰ τῶν βαρβάρων. | μετὰ τὸν θεὸν οὐδίναν | ἔχομεν ἡμῆς βοηθὸν | ὑμῶν. ἔπεμψά σου οὖν, | κύριέ μου, τὸν Ἀθιόειν·| ἀξιῶ τοὺς

As it transpires from the text, the mother of Moses the soldier, whose name is unknown to us, decided to write to the *praepositus Flavius Abinnaeus*, commander of the garrison at Dyonisias (northwest corner of the Fayyum) to request five days' leave for her son.[26] She takes this opportunity to notify him that she also sent her other son Heron to serve in the army, but it seems that he joined some presumably dissident troops.[27] Moses' name makes it highly likely that he was a Jewish soldier.[28] In this regard, it is worth pointing out that his brother's name, Heron, was also in use among Jews and Samaritans.[29] Finally, the providentialist expression "after God" (μετὰ τὸν θεὸν) also points in this direction.

Beyond the issue of names as a marker of identity, considering that, in Late Antiquity, Christians chose Biblical names even more than Jews themselves, it is almost impossible to determine the Christian or Jewish nature of the Biblical names of the soldiers appearing in two papyri dated to the 6th century CE.[30] The first one—dated to 507 CE—features the brothers Aurelius Isaac, Flavius Silvanus and Flavius Serapion, the last two being soldiers in the *numerus* of Hermopolis. According to the text, their sister Arconcia was married to Flavius Bes Nikon, another soldier serving in the same unit.[31] Most of the soldiers who appear in the second papyrus—dated to 573 CE—were Christians, as the famous *chi-ro* monogram appears in the heading. Nevertheless, we may hypothesise that some of them, who bore names that were prevalent among Jews, may have been Jewish: Flavius Jacob (drummer), son of Isaac and Tlous; Flavius Abraham (*actuarius*), son of Stercorius, or Flavius Abraham (centurion), son of Pamet; all of whom belonged to the legions of Syene and Philae.[32]

1.3 *The Western Diaspora*

As is well known, the Jewish catacombs in Rome have provided us with a wealth of funerary inscriptions. The Vigna Randanini catacomb has yielded an unusual

πόδας σου, | κύριε πάτρωνι, ἐπιδὴ | οἴδης καὶ σὺ ὅτι ἡ πέντη | ἡμέραι ὁ ἐνιαυτὸς ὅλος | εἰσί, ἀξιῶ σε καὶ παρακαλῶ σε, | κύριε, ἵνα συνχωρήσῃς | αὐτοῦ τὰς ὀλίγας ἡμέρας. | ἤ τι κελεύεις, κύρ[ι]ε, εἶπον | τοῦ Ἀθιόει. | ἐρρῶσθαί σε εὔχομαι, | κύριε πάτρωνι. Cf. Bagnall and Cribiore 2006, 358–359.

26 Perea Yébenes 2020, 265–267.

27 This is the sense Sabino Perea Yébenes (2020, 266) gives to the expression τῶν βαρβάρων. Conversely, the editor takes this to mean joining the military.

28 Ilan 2008, 290. Cf. Mussies 1994, 243.

29 Ilan 2008, 136–138.

30 Varon 2001, 276–277.

31 Hunt and Edgar 1932–1934, I, 61 (= *P.Lond*, 992).

32 Hunt and Edgar 1932–1934, I, 30 (= *P.Lond*, 1722). See Varon 2001, 277. On the name Flavius in the field of the military, see Keenan 1973.

Latin inscription written in Greek characters—dated to the 3rd or 4th centuries CE—dedicated by Reginus to his colleague Agrius Evangelus:

> For Agrius Evangelus, well-deserving, Reginus his colleague (had this made).[33]

The meaning of this inscription hinges on the meaning granted to the term κολληγα. As in Christian inscriptions, *collega* was used to indicate a fellow-soldier,[34] and we should therefore conclude that Reginus, the dedicant of the epitaph, had been a comrade in arms of Agrius Evangelus.[35]

Along with Rufinus' inscription (Appendix II), the epigraphic evidence from the Jewish catacombs of Rome suggests that the capital of the Empire became an attractive destination for retired Jewish soldiers who had completed their military service. One curious rabbinic tradition might confirm this. In a Babylonian tradition a certain, otherwise unknown sage, Hanan son of Tahlifa, reports to Rav Josef (d. 320 CE)[36] that he once met someone who owned an ancient scroll written in Hebrew. This antiquarian told the rabbi that he had been a soldier in the Roman army and had found—and apparently acquired—the scroll in the archives of Rome (גינזי רומי). It seems that the scroll contained certain apocalyptic passages describing various messianic wars at the end of time, including chronological calculations and theological speculations:

> R. Hanan b. Tahlifa sent [word] to R. Joseph: I once met a man who possessed a scroll written in Hebrew in Assyrian characters [Aramaic writing]. I said to him: 'Whence has this come to thee?' He replied, 'I hired myself as a mercenary in the Roman army, and found it amongst the Roman archives. In it is stated that four thousand, two hundred and thirty-one years after the creation the world will be orphaned. [As to the years following,] some of them will be spent in the war of the great sea monsters, and some in the war of Gog and Magog, and the remaining [period] will be the Messianic era, whilst the Holy One, blessed be He, will renew his world only after seven thousand years.' [...].[37]

---

33 *JIWE 2*, No. 264= *CIJ I*, No. 207: Ἀγρι Εὐαν|γε⟨λ⟩ο βενεμε|ρεντι, Ῥηγεινους | vac. κολ⟨λ⟩ηγα. vac. (*Agrio Evangelo benemerenti, Reginus collega*).

34 For instance, *ILCV*, 448. See *JIWE 2*, 231.

35 Schoenfeld 2006, 123.

36 According to the main authority on rabbinic chronology, Rav Sherira Gaon. See Rabinowitz 1988, 102.

37 *BT*, *Sanhedrin*, 97b. See Ilan 2009, 367.

We must keep in mind that this story belongs to the late legendary traditions of the so-called "Jewish treasures in Rome" transmitted only by Babylonian rabbis, and that its historical value is therefore limited. Nevertheless, it contains some unique features that set it apart from the common *topoi* on the Jewish treasures of Rome. It is certainly surprising that it features a Jew who had served as a soldier in the Roman army, implying the knowledge and acquiescence of rabbis to this sort of "assimilation". As Tal Ilan notes, "this is an incidental reference of Jewish accommodation to Rome and therefore may record a known fact. This Jewish soldier had found a scroll written in Hebrew characters in Rome. Yet it is not a Torah scroll. Instead, it tells of messianic wars, the description of which fits well with the sort of apocalyptic literature that Jews produced during the Second Temple period. If there is a grain of truth in it, it is further evidence to the kind of Jewish libraries and Jewish literature stored in Rome that an antiquarian or scholar of Judaism could have found there."[38] It is obvious that this soldier was a Jew: if he was not, why would he have been interested in a Hebrew scroll, and how would he have had access to it? We must also keep in mind that the soldier was able to explain the apocalyptic contents of the scroll, which shows that he had read it.

Finally, two fourth-century inscriptions found in Oescus (modern-day Gigen), in Moesia Inferior (Bulgaria), the home-base of *Legio V Macedonica*,[39] list the names of *Flavius Euphorbius*—from Phocaea, Asia Minor—and *Flavius Valentianus*—from Syria-Palaestina—who were *primipularii* [*primipilarii*] of this legion.[40] The same location has yielded another epitaph featuring the name of *Sanbatio* but, despite its Hebrew origins, it is not clear whether the name is used in a Jewish or Christian context.[41] Although we cannot know for certain if these soldiers were Jewish,[42] it remains possible that at least one of them was a Jew, or that this long-lived military camp still had Jewish soldiers at the time, especially considering that a fragmentary Greek inscription with a menorah, dated to the the 4th or 5th centuries CE, was found there.[43]

38 Ilan 2009, 368.

39 Rodríguez González 2001, I, 195–207.

40 Beševliev 1952, nos. 77–78.

41 *ILBulg* 119. See Panayotov 2004, 62.

42 Alexander Panayotov (2004, 62) writes: "Whether there were Jews enlisted in the Moesian legions or not remains, however, an open question."

43 *IJO I*, Moes2.

2 Under the Christian Empire

2.1 *The Tower of the Jews*

In his *On buildings*, 6th-century historian Procopius of Caesarea mentions the existence of a tower known as Ἰουδαῖος (Judaeus) among the fortifications on the Danube border, in Dacia Ripensis (part of the former province of Moesia Superior) which, being in ruins, was rebuilt by emperor Justinian (527–565 CE):

> And he [Justinian] remodelled a stronghold called Judaeus, which had consisted of a single tower, and made it a splendid fortress in name and in fact.[44]

It seems that it was a watchtower (πύργος, *turris*) built on or near the military road which ran from *Singidunum* (Belgrade), through *Viminacium* (Kostolac), *Bononia* (Vidin), *Ratiaria* (Arčar), *Oescus* (Gigen), *Novae* (near Svishtov), and *Durostorum* (Silistra) to the Danube delta.[45] According to Procopius, the tower was turned into a φρούριον (Latin *castellum*), i.e. a bigger fortification. Vegetius uses this term to denote a small fortified settlement or garrison of an auxiliary unit.[46]

Veselin Beševliev made a convincing point that the Greek name of the tower was, in fact, the transcription of a Latin ablative with a locative meaning—*Iudaees*, i.e. *Iudaeis*—and thus that the reading of the manuscripts is correct. In fact, there are similar transcriptions of the Latin names quoted by Procopius in his lists of fortifications.[47] Thus, the name of this watchtower should be translated as "At the Jews". Examples of names of *stationes* that indicated the ethnic origins of their occupants or of those who first settled there are available both in Procopius and in the Roman itineraries.[48] In the city territory of Remesiana (modern-day Byala Palanka) there was a stronghold named, apparently after a road-station, Γραῖκος, i.e. *ad Graecos*.[49] The *Tabula Peutingeriana* features a road-station with a similar name in Italy[50] and also lists the road-station *ad*

44 Procopius of Caesarea, *De aedificiis*, IV, 6, 21: καὶ πύργον ἕνα, Ἰουδαῖος καλούμενον, φρούριον διεσκευάσατο κεκαλλιστευμένον καλεῖσθαί ⟨τε⟩ καὶ εἶναι.

45 Panayotov 2004, 52.

46 Vegetius, *Epitoma rei militaris*, III, 8, 21–22.

47 For example, the names of the following forts in *De aedificiis*, IV, 4: Βαρβαρίες (Barbaris) Ἑταίρες (Haeteris); Δουλίαρες (Doliaris). See Beševliev 1967, 103–104.

48 On this point, I closely follow Panayotov 2004, 54. Cf. Tibor 2016, 14. On the predominantly military nature of *stationes*, see *supra*, p. 82 note 248.

49 Procopius of Caesarea, *De aed.*, IV, 4.

50 *Tabula Peutingeriana*, *segmentum* IV, 4.

Germanos in North Africa.[51] In Epirus Nova there were forts called Ἡδονία and Ἥδωνες.[52] The fort Σαρμαθών appears in Haemimons,[53] and there is a road-station designated in the sources as Σάρματες.[54] Finally the *castelli* Σκυθίας and Βαστέρνας are found in the provinces of Scythia and Moesia Secunda.[55]

Procopius' brief description of this Ἰουδαῖος tower, which was probably located somewhere on the road between *Dorticum* and *Bononia*, does not allow us to date it accurately, but it seems obvious that it existed prior to when Justinian began to rebuild the fortresses on the Danube.[56] Nevertheless, we may narrow down the time when this road-station was established there based on the constitution of Valentinian and Valens from 364 to Tautomedes, ruler of Dacia Ripensis, who ordered the construction and repair of watchtowers in the province:

> On the border entrusted to Your Gravity, you shall, during the time of your administration, annually construct towers in suitable places, in addition to the towers which must be repaired, if any, perchance, need repair [...].[57]

Taking this date as the *terminus ante quem*—the text refers to repairs on previously-existing towers—, Alexander Panayotov suggested that the original construction of the tower of the Jews dates back to an earlier period as far back as the Severan dynasty, when some Jews came to Pannonia, although we lack the evidence to make a categorical statement on the subject.[58]

2.2 *The Attack on the Church of Theonas*

Shortly after the defeat of the usurper Magnentius in 353, Constantius II, an Arian Christian emperor, consolidated the Empire under his rule. However, from an ideological point of view, the controversial figure of Athanasius, the influential Catholic bishop of Alexandria, threatened the unity Constantius II had managed to impose on the Church and the State. It was not long before dialectic confrontations gave way to violent action from which the Catholic bishop of Alexandria managed to escape with his life, albeit at the cost of suf-

51 *Tabula Peutingeriana, segmentum* IV, 2.
52 Procopius, *De aed.*, IV, 4. See Haury 1965, 117 and 118.
53 Procopius, *De aed.*, IV, 11. See Haury 1965, 147.
54 Procopius, *De aed.*, IV, 4. See Haury 1965, 122.
55 Procopius, *De aed.*, IV, 11. See Haury 1965, 148.
56 Panayotov 2004, 56.
57 *CTh.*, 15.1.13: *In limite gravitati tuae commisso praeter eas turres, quas refici oportet, si forte indigeant refectione, turres administrationis tempore quotannis locis opportunis extrue.* [...]
58 Panayotov 2004, 56.

fering a third exile that would only end after the death of Constantius II, six years later. A passage from Athanasius' defence, which was sent to emperor Constantius II by bishop Lucifer of Cagliari, reveals that a Jewish military unit led by the *dux Aegypti Syrianus*[59] participated in the violent assault against the church of *Theonas*, in Alexandria, on the night of 8–9 February 356. Lucifer tells of how certain Jewish soldiers broke into the *domus dei* intending to commit a massacre:

> [...] If you [emperor Constantius] deny that you are among those to whom the Lord says: "you serpents, generation of vipers!" prove that it was not you, as a soldier, but only the Jews sent to Alexandria, only the army of the Jews that forced the gates of the house of God, only Syrianus, who was the commander of the Jewish soldiers; prove that it was only the Jews who barged into the basilica bearing arms and who murdered a great multitude [...].[60]

Although he does not specify the composition of the troops, Athanasius of Alexandria states that he himself was gathered with his flock holding a vigil when the heinous events took place:

> We have already protested against the nocturnal assault which was committed upon ourselves and the Lord's house; although in truth there needed no protest in respect to proceedings with which the whole city has been already made acquainted. For the bodies of the slain which were discovered were exposed in public, and the bows and arrows and other arms found in the Lord's house loudly proclaim the iniquity. But whereas after our Protest already made, the most illustrious Duke Syrianus endeavours to force all men to agree with him, as though no tumult had been made, nor any had perished (wherein is no small proof that these things were not done according to the wishes of the most gracious Emperor Augustus Constantius; for he would not have been so much afraid of the consequences of this transaction, had he acted therein by

59 On Syrianus, see Jones, Martindale and Morris 1971 (*PLRE I*), 872 (*Syrianus*).

60 Lucifer of Cagliari, *De Athanasio*, II, 22: [...] *Aut si negas esse te ex illis qui audiunt ore domini: serpentes, generamina uiperarum, proba non te, sed Iudaeos destinasse militem ad Alexandriam, Iudaeorum militem obsedisse fores dei domus, Iudaeorum militum decem fuisse Syrianum, proba Iudaeos ingressos basilicam cum armis atque certum numerum interfecisse* [...].

> command); and whereas also, when we went to him, and requested him not to do violence to any, nor to deny what had taken place, he ordered us, being Christians, to be beaten with clubs; thereby again giving proof of the nocturnal assault which has been directed against the Church. [...][61]

According to his account, there can be no doubt that the attack on the church was the result of a perfectly planned military operation rather than a violent riot by a Jewish or Greek mob.[62] Thus, Lucifer's information, which certainly provides more details with regards to the composition of the military force that participated in the events, does not contradict Athanasius' account of the incident.

Without substantiating his claims, David Wood believed that the Jewish military unit in question was the *Regii Emeseni Iudaei*,[63] a "phantom" unit whose existence is entirely based on a mistaken reading from the military necropolis of Concordia (northern Italy).[64] We are, in fact, ignorant of the nature of this *numerus militum Iudaeorum* under the command of the *dux Aegypti Syrianus* (*Iudaeorum militum ducem fuisse Syrianum*), but it seems unlikely that it would have moved from outside Egypt to join the forces that seized the city of Alexandria as a result of the religious and political conflict between emperor Constantius and the influential bishop of Alexandria.[65] Ralf Scharft may be cor-

61 Athanasius, *Historia Arianorum ad monachos*, 81, 2–4: Ἤδη μὲν διεμαρτυράμεθα περὶ ἧς πεπόνθομεν νυκτερινῆς ἐφόδου ἡμεῖς τε καὶ τὸ κυριακόν, εἰ καὶ διαμαρτυρίας χρεία οὐκ ἦν ἐφ' οἷς πῶσα ἡ πόλις ἔγνακέ τε καὶ γινώσκει. Τά τε γὰρ εὑρεθέντα σώματα τῶν ἀναιρεθέντων δημοσίᾳ προετέθη καὶ τὰ ἐν τῷ κυριακῷ ὅπλα τε καὶ τόξα κέκραγε τὴν παρανομίαν. ἐπειδὴ δὲ καὶ μετὰ τὴν διαμαρτυρίαν ὁ λαμπρότατος δοὺξ Συριανὸς βιάζεται πάντας συνθέσθαι αὐτῷ, ὡς οὔτε θορύβου γενομένου οὔτε τινὸς ἀποθανόντος, ἔλεγχος δὲ οὗτος οὐκ ὀλίγος μὴ γεγενῆσθαι ταῦτα κατὰ γνώμην τοῦ φιλανθρωποτάτου Αὐγούστου Κωνσταντίου. Οὐκ ἂν γὰρ ἐφοβήθη ἐπὶ τοῖς οὕτω γενομένοις, εἰ ἐκ προστάξεως ταῦτα πεποιήκει. Καὶ γὰρ καὶ ἀπελθόντας ἡμᾶς πρὸς αὐτὸν καὶ ἀξιοῦντας μηδενὶ βίαν ποιεῖν μηδὲ ἀρνεῖσθαι τὰ γενόμενα ἐκέλευσε Χριστιανοὺς ἡμᾶς ὄντας κατακοπῆναι ῥοπάλοις δεικνὺς καὶ ἐκ τούτων τὸν γενόμενον νυκτὸς πόλεμον κατὰ τῆς ἐκκλησίας. [...] Cf. *Apologia pro fuga sua*, 24; *Apologia ad Constantium imperatorem*, 25. In the *Historia acephala* (5, 1–14) it is written that the attack on the church took place exactly on the night of 8–9 February 356. See Schoenfeld 2006, 123; Gwynn 2007, 162–164 and *Idem* 2012, 43–44; Olshanetsky 2018, 17–18.

62 It is remarkable that Athanasius (*Hist. Arian. ad mon.*, 81, 5) mentions the *agentes in rebus* (κουριώσοι) involved in this incident, both as informants and witnesses in the subsequent trial (Di Paola 2005, 67). It is worth keeping in mind that there were Jews in the *schola agentum in rebus*.

63 Wood 1992.

64 See Appendix III.

65 Barnes 1993, 115–119 and 129–130.

rect in his identification of these Jewish soldiers, whom Athanasius described as legionaries (λεγεώνων στρατιωτῶν),[66] as members of the regular Roman army in Egypt.[67]

2.3 *An Apparently Attractive Career*

We do not know whether the well-known legal tradition that exempted Jews from the rites and oaths of the military gave rise to friction within the army over time, or if commanders failed to understand the reasons for which Jews had been granted a number of exceptional rights, perhaps due to the fact that they perceived them as unfair treatment. Be that as it may, it seems that these concessions were generally respected by later emperors. In fact, their traditional validity was actively safeguarded during the Diocletianic persecutions against Christians. Thus, there is a rabbinic tradition that reflects a general decree by Diocletian in which it is stated, to prevent any misunderstandings, that Jews were exempted from offering all manner of libations.[68] The unambiguous presence of Jewish identity within the military sphere can be taken as proof of the fact that, in practice, the necessary conditions were in place for the service of Jews in the imperial army. In previous centuries, military service had provided Jews with a comfortable means of escaping other burdensome duties, and it likewise provided those on the lower rungs of the social ladder, including those who had been slaves and gladiators, with a way of life.[69] Indeed, there was no shortage of reasons to enlist in the army, most of which were quite attractive. Apart from pay (the immediate goal of those who lived in poverty) and the prospect of loot and pillage during war, soldiers in the *auxilia* could attain Roman citizenship as a discharge bonus when they retired, a right they could pass on to their offspring. For the state, this right was an effective means of integrating these soldiers and their families into the Roman social order which,

---

66 Athanasius, *Hist. Arian. ad mon.*, 81, 7.

67 Scharft 1997, 347.

68 *JT*, *'Abodah Zarah*, 5:4, 44d: [...] when King Diocletian came here he decided and said, all peoples have to offer libations except the Jews [...]. See Avi-Yonah 1984, 45. See on this subject Stemberger 2000, 24; Hezser 2011, 280. Much has been said about whether the obligation to offer libations should be dated to the period of Diocletian's rule or to the anti-Christian persecutions carried out by Maximian in 307 CE. On this subject, see Rabello 1984, 150–151, who dates the decree back to 303. As epigraphy shows, the full integration of Jews into the civic sphere of Roman cities was only possible if there were no formal impediments against to their religious customs. For example, we know that, in the fourth century, the Jews of Cyme (Asia Minor) described themselves as citizens of that city. See Lafli 2021.

69 Hild 1884–1885, 29.

in turn, could lead to the consolidation of a sort of patriotic feeling of unity among the troops, although this aspect, being subjective, is very hard to assess properly.[70] Moreover, soldiers were exempted from the poll tax or *capitatio*, and during the 4th century this exemption was extended to their closest relatives or, if they were without kin, it could be compensated for with deductions from their taxes on land.[71] By virtue of their *immunitas*, veterans likewise enjoyed other prerogatives, such as exemption from duty, from *collatio lustralis* (a tax on goods and services) and, especially, from curial taxes.[72]

Though an exhausting and always risky endeavour, it seems, according to a sermon by Augustine of Hippo recently discovered by François Dolbeau, military service in the Roman army did not lose its lustre under the Christian Empire, as the prospect of a comfortable life for veterans remained a magnificent stimulus:

> [...] Consider the work of the soldiers who bear arms, the perils they face, the hardships and adversities they bear, in the heat, in the cold, enduring hunger, thirst, injury and death. They face danger on a daily basis, yet they cast not their gaze on the toil of the soldier but on the leisure of the veteran. "Here," they say, "work will end, and in but a few years we will have respite, life will be good, we will spare no expense, we shall have immunity, we will not be called to municipal offices, and we shall bear no-one's burden once we have borne that of soldiery." [...].[73]

Insofar as military service remained attractive in certain sectors of society throughout the 4th century, Jews also shared those aspirations by continuing to enlist in the army, at least until the Christian legislation of the early 5th century barred them from doing so.[74] A homiletic *Midrash* has preserved the memory

70 See Campbell 2002, 34–35; Roux 2020, 27–32. Cf. Cheesman 1914, 34–35.

71 Jones 1992, 617 and 675.

72 Jones 1992, 618–619, 635–636 and 646–649; Speidel 1992; Southern and Dixon 2000, 87–88. On the tax benefits of retired soldiers, see Chastagnol 1977. On testamentary privileges and the *peculium castrense*, see Campbell, 1984, 210–236. For a general vision of the legal status of soldiers and veterans, see Alston 1995, 53–68.

73 Augustine of Hippo, *Serm. Dolbeau*, 15, 4 = *Serm. Mayence*, 45, 4: [...] *Considerate labores militantium, arma portantium, quae pericula subeant, quam dura et ardua tolerant, in frigoribus, in aestubus, fame, siti, uulneribus, mortibus. Cottidie uersantur in periculis, non sibi proponentes ante oculos laborem militis, sed otium ueterani. "Ecce, inquiunt, finitur labor, post paucos annos succedet otium, bene nobis erit, sumptus non deerunt, immunitas aderit, ad nullas functiones ciuitatum deuocabimur, nemo nobis post militiae sarcinam imponet suam."* [...]

74 Juster 1914, II, 277.

of an anecdote, which seems to originate in Late Antiquity, according to which a young Jew's parents attempted to convince the officer in charge of recruitment to admit him to the army, insisting that he met the standards required for military service:

> [...] It is like the conscription of soldiers. Someone came to conscript one man's son. He said to him: Look at my son, what a strong young man he is, and how tall! Even his mother said: Look how tall our son is! He said to them: in your eyes he is strong and tall. I don't know if he really looks so tall in my eyes. They measured him, found him too short and excluded him. [...][75]

A very similar case, collected in the texts of the *Midrash Genesis Rabbah*, was also a matter of discussion among wise men:

> Rabbi Elazar beRabbi Shimeon said: [It is like] a state which was conscripting for an *'inqamon* [tall bodyguard] for the king, and there was one woman there who had a son who was a dwarf, whom she called 'Makroelaphros' [tall and fleet of foot]. She said 'My son is tall and fleet of foot, why don't you conscript him?' They said to her 'In your eyes he is tall and fleet of foot, in ours he's a dwarf!'[76]

In fact, it is remarkable that both texts perfectly agree with the understandable Roman demand that those who were conscripted to the army should be of a certain known minimum height, as Vegetius writes:

> At what height recruits should be approved. The height of recruits was, I know, always required to be up to the *incomma*, so that men of 6 ft. (= 5 ft. 9½ in., 1 77 m.) or at least 5 ft. 10 in. (= 5 ft. 7½ in., 1 72 m.) were approved for the *alares* cavalry or the First cohorts of the legions. [...][77]

Although compulsory and hereditary conscription was established in the Empire under Diocletian, there was the option to avoid military service by

---

75 *Aggadat Bereshit* XL. See Oppenheimer 2005a, 188. Cf. Isaac 1990, 302–303.

76 *Midrash Genesis Rabbah*, LXV. See Oppenheimer 2005a, 189.

77 Vegetius, *Epitoma rei militaris*, I, 5: *Qua statura iuniores probandi sint. Proceritatem tironum ad incommam scio semper exactam ita, ut VI pedum uel certe V et X unciarum inter alares equites uel in primis legionum cohortibus probarentur.* [...] See Oppenheimer 2005a, 189.

paying a tax known as *aurum tironicum*.[78] It is very likely that wealthy Jewish families made use of this to exempt their sons of military age from serving in the military. As Aharon Oppenheimer writes, "it seems surprising at first that the midrashic source should present parents asking for their son to be conscripted and being proud of his height. This phenomenon does not seem to square with the information about Jews who, on the contrary, wanted to free their sons from the obligation to serve in the Roman army by paying for their release. However, it is certainly possible that there were parents who wanted their son to serve in the army, for soldiers received high wages and together with their families were exempt from a number of different financial dues."[79]

It is true that there is a dearth of direct information on the service of Jews in the Late Roman army. We know that, with the exception of Illyrian units, *limitanei* (border troops) were generally drawn from the local populations. It is quite possible that Jews continued to serve as frontier soldiers at the time, as sources point to the fact that, in this specific case, they were compulsorily subjected to local conscription.[80] In spite of this, some of them managed to carve out careers in the Roman army and, through their years of service and their achievements, to pursue the *cursus honorum*. This seems to have been the case of Benjamin, a *centenarius* of the troops stationed in the *limes*.[81] Nevertheless, other Jews (perhaps the majority) preferred to substitute the payment of the so-called *aurum tironicum* in lieu of being drafted.

According to a rabbinic source, at the time of Rabbi Mana—second half of the fourth century—a *nimorah* (= Latin *numerus*, i.e. Roman cohort) stationed in Sepphoris wanted to take the sons of the inhabitants of the city away with them. Rabbi Mana is said to have made a proclamation that the inhabitants should pledge their houses to redeem their sons with the proceeds. In other words, parents would pay the conscripts' tax (*aurum tironicum*) in order to free their sons from military service:

> In the days of Rebbi Mana there was a company in Sepphoris, and their children were taken as pledges. When they were about to leave, Rebbi Mana published (so) [a proclamation] following Rebbi Immi. He said, not

78 Ammianus Marcellinus, *Res gestae*, XIX, 11, 7. See Rostovtzeff 1918; Jones 1992, I, 152, 434 and 615; Southern and Dixon 2000, 67; Stemberger 2000, 12 and 281; Hebblewhite 2017, 99 and 133.

79 Oppenheimer 2005a, 189.

80 Kasher 1985, 78; Schoenfeld 2006, 123; Olshanetsky 2018, 23.

81 Stemberger 2000, 306.

> that I am of his opinion, but because of the Sepphoreans that they should not permanently lose their children.[82]

Actually, all signs point to the fact that, throughout the 4th century, many Jews enlisted voluntarily out of a wish to reap the benefits provided by the military.

2.4 *The Church Hierarchy and the Exclusion of Jews from the Roman Army*

Jews began to enlist in the army in ever-greater numbers, perhaps due to the fact that, at the time, a career in the military had many more benefits than drawbacks. Some Christian authors, such as Sulpicius Severus, voiced their concern about the great number of Jews who had "infiltrated" the army:

> [...] And it is also evident that barbarous nations, and especially Jews, have been commingled with our armies, cities, and provinces; and we thus behold them living among us, yet by no means agreeing to adopt our customs.[83]

According to this account, they must not have faced any significant difficulties when it came to entering the *militia armata*.[84] In fact, pagan sacrifices, ceremonies and festivities or the imperial cult ceased to be—if indeed they ever were one—a problem under the Christian Empire,[85] and it would seem that the new *sacramentum militiae* through which soldiers consecrated their lives to the military profession, was not an insurmountable obstacle.[86] As José Fernández Ubiña has pointed out, the notion of "Christianisation" of the army under the Later Roman Empire deserves some nuance. Indeed, contemporary Christian sources reveal that it was not uncommon to justify obedience towards unbelieving or heretical commanders as "serving God", or acknowledging that the only law by which troops were governed was the emperor's will.[87] According to

---

82 JT, *Pesaḥim*, 4:9, 31b. See Isaac 1990, 302–304; *Idem* 1991; Oppenheimer 2005, 184; Hezser 2011, 280.

83 Sulpicius Severus, *Chron.* (ca. 403), II, 3, 6: [...] *siquidem Romanum solum ab exteris gentibus aut rebellibus occupatum aut dedentibus se per semper pacis speciem traditum constat, exercitibusque nostris, urbibus atque prouinciis permixtas barbaras nationes, et preaecipue Iudaeos, inter nos degere nec tamen in mores nostros transire uideamus*. See Juster 1914, II, 276; Applebaum 1971, 182.

84 Delmaire, Rougé and Richard 2005, 407 note 3.

85 Letta 2020, 159–162.

86 With regards to baptism as a *sacramentum* of the *militia Christi*, see Zuccotti 2000, 92.

87 Fernández Ubiña 2000, 589, who also aptly points out that that, for this reason, Arians and

this scholar, we should consider that "in fact, religious differences remain the dominant theme in the most reliable historical texts on Late Roman armies. Even in the East, where there was always a considerably greater number of Christians, the documents preserved in the *Archives of Abinnaeus*, the pagan commander of the Egyptian fortress of Dionysias (El Fayyum), reveal that religious beliefs had no effect on the daily lives of the soldiers or their relations with the civilian population."[88]

Nevertheless, in a period dominated by Christian dogma and a progressive anti-Jewish ideology,[89] both the Empire and the Church were reluctant to accept this reality.[90] In fact, the Church Fathers could not help being wary of the common presence of Jews (some of them even serving as commanding officers) in the "Christian armies", nor could they tolerate that they should hold important offices within the public administration, for theological reasons.[91] Even some pagans, such as Claudius Rutilius Namatianus, who was also susceptible to anti-Jewish ideas, were very close to the thought of the Christian author Sulpicius Severus. Indeed, many shared the Gallic poet's words:

> [...] If only the Jewish state had never been subdued by Pompey's wars and Titus's command! The sickness of this plague, though cauterized, spread far; a conquered nation rules its conquerors. [...].[92]

In spite of the official acknowledgement of Judaism as *licita religio*, which was inherited from the Roman legal tradition, imperial power would ultimately

---

other heretics in the Roman army were immune to anti-heretical legislation. As this author has proved, it does not seem that Christians themselves had any qualms about offering sacrifices to the idols during the pagan period of the Empire, and Roman authorities do not seem to have pressed the matter except in periods of persecution (Fernández Ubiña 2000, 427). The Roman army was always governed by a certain code of tolerance towards different religious practices and beliefs. "In fact, Rome did not impose any specific religious policies in the army, nor did it take the least interest in the personal beliefs of soldiers or any proselytizing activities they may have engaged in" (Fernández Ubiña 2000, 598).

88 Fernández Ubiña 2000, 590. In this regard, it is logical to assume that, if there was a large number of pagan schools of rhetoric to teach State civil servants under Julian (when the Church was banned from teaching rhetoric by law), pagans and Jews must have continued to hold a majority of positions in the administration and the army after the so-called "triumph of the Church" (see Stemberger 2000, 191–192).

89 González Salinero 2000.

90 See Whelan 2018, 79.

91 On this subject, see González Salinero 2000, 104 ff.

92 Rutilius Namatianus, *De reditu*, I, 395–398: [...] *Atque utinam numquam Iudaea subacta fuisset / Pompeii bellis imperioque Titi! / Latius excisae pestis contagia serpunt / Victoresque suos natio victi premit* [...].

give in to anti-Jewish ecclesiastic pressure. Thus, in the year 404 CE, Honorius denied Jews (and Samaritans), who had hitherto enjoyed the privileges granted to the *agentes in rebus*, the right to belong to the *militia*:

> THE SAME TWO AUGUSTI [ARCADIUS AND HONORIUS]. TO ROMULIANUS, PRAEFECTUS PRAETORIO. We order that the Jews and the Samaritans, who delude themselves with the privilege of the Executive Agents, shall be deprived of any State office. GIVEN ON THE TENTH DAY BEFORE THE CALENDS OF MAY AT ROME, IN THE CONSULATE OF HONORIUS AUGUSTUS FOR THE SIXTH TIME AND OF ARISTAENETUS.[93]

According to Vittore Colorni's interpretation, Jews were excluded from the army for the first time pursuant to this law;[94] however, this is not entirely correct. In this regard, it is worth pointing out that, at the time, the term *militia* had an even broader sense: it was used to designate service in the imperial administration writ large, in official positions of both a civil and military nature.[95] Nevertheless, it is also worth pointing out that the *agentes in rebus*—a kind of official in charge of the intelligence service of the Roman state—were an imperial corps that always had a military nature since its creation in the early 4th century.[96]

A few years later, in 418 CE, the same emperor banned Jews once again from entering imperial service (*militia*), allowing those who had been sworn in as *agentes in rebus* or *palatini militiae* to carry out their duties during the allotted period of time. However, those who were serving in the army (*militia armata*) were compelled to abandon it forthwith, as this new constitution established that they would thereafter be banned from it and that none of their former merits could be brought to bear against this decision:

---

93 *CTh.*, 16.8.16 (Linder 1987, No. 33, 223): *IDEM AA, ROMULIANO P(RAEFECTO) P(RAETORI)O. Iudaeos et Samaritanos, qui sibi agentum in rebus privilegio blandiuntur, omni militia privandos esse censemus. DAT. X KAL. MAI. ROM(AE) HONO(RIO) A. VI ET ARIST(AENETO) CONSS* (22 April 404). See Juster 1914, II, 244 and 277; Demougeot 1960, 278; Reichardt 1978, 36; Rabello 1987–1988, II, 696; Vogler 1979, 57 and 67; Lucas 1993, 65; Noethlichs 1996, 106; Jordán Montes 1997, 125; De Bonfils 1998, 117–118; González Salinero 2000, 111; Noethlichs 2001, 127; Varon 2001, 276; Magnou-Nortier and Rouche 2002, 340–341; Delmaire, Rougé and Richard 2005, 390–393; Roth 2007, 419; Nemo-Pekelman 2008, 417–419; *Eadem* 2010, 172, 178, 183–194; Kraemer 2020, 177, 237.

94 Colorni 1956, 24.

95 Linder 1987, 223.

96 See *supra*, p. 103 note 8.

THE SAME TWO AUGUSTI [Honorius and Theodosius II] TO PALLADIUS [*Praefectus Praetorio* of Italy from 415 (or 416) to 421], *PRAEFECTUS PRAETORIO*. The entrance to the State Service shall be closed from now on to those living in the Jewish superstition who attempt to enter it. We concede therefore to all those who took the oath of the Service, either among the Executive Agents or among the Palatins, the opportunity to terminate their service on its statutory term, suffering the deed rather than encouraging it, though what we wish to be alleviated at present to a few shall not be permitted in the future. As for those, however, who are subject to the perversity of this nation and are proven to have entered the Military Service, we decree that their military belt shall be undone without any hesitation, and that they shall not derive any help or protection from their former merits. Nevertheless, we do not exclude Jews educated in the liberal studies from the freedom of practicing as advocates, and we permit them to enjoy the honour of the curial liturgies, which they possess by right of their birth's prerogative and their family's splendor. Since they ought to be satisfied with these, they should not consider the interdiction concerning the State Service as a mark of infamy. GIVEN ON THE SIXTH DAY BEFORE THE IDES OF MARCH AT RAVENNA, IN THE CONSULATE OF THE TWO AUGUSTI, HONORIUS FOR THE TWELFTH TIME AND THEODOSIUS FOR THE EIGHTH.[97]

97 *CTh*., 16.8.24 (Linder, 1987, No. 45, 281): *IDEM AA. PALLADIO P(RAEFECTO) P(RAETORI)O. In Iudaica superstitione viventibus adtemptandae de cetero militiae aditus obstruatur. Quicumque igitur vel inter agentes in rebus vel inter palatinos militiae sacramenta sortiti sunt, percurrendae eius et legitimis stipendiis terminandae remittimus facultatem, ignoscentes facto potius quam faventes, in posterum vero non liceat quod in praesenti paucis volumus relaxari. Illos autem, qui gentis huius perversitati devincti armatam probantur adpetisse militiam, absolvi cingulo sine ambiguitate decernimus, nullo veterum meritorum patrocinante suffragio. Sane Iudaeis liberalibus studiis institutis exercendae advocationis non intercludimus libertatem et uti eos curialium munerum honore permittimus, quem praerogativa natalium et splendore familiae sortiuntur. Quibus cum debeant ista sufficere, interdictam militiam pro nota non debent aestimare. DAT. VI ID. MART. RAV(ENNAE) HONOR(IO) XII ET THEOD(OSIO) VIII AA. CONSS* (10 March 418). See. Juster 1914, II, 278; Demougeot, 1960, 279; Reichardt 1978, 36; Vogler 1979, 57; Linder 1987, 76 and 280–283; Rabello 1987–1988, II, 696; Jones, 1992, 948; Lucas 1993, 65; Noethlichs 1996, 108; Jordán Montes 1997, 125; De Bonfils 1998, 194–196; Williams 1998b, 106; González Salinero 2000, 112; Noethlichs 2001, 127; Varon 2001, 276; Magnou-Nortier and Rouche 2002, 350–353; Schoenfeld 2006, 125; Lee 2007, 190; Delmaire, Rougé and Richard 2005, 406–409; Roth 2007, 419; Nemo-Pekelman 2008, 417–420; *Eadem* 2010, 172–173, 178, 183–194; Kraemer 2020, 177–178, 237, 250–251, 270. We must remember that pagans had already been excluded from the army and the imperial

There can be no doubt that, in spite of the emperor's scornful generosity in what concerns the legal profession or the acknowledgement of curial service (which, moreover, was perceived as a burdensome disadvantage rather than an admirable honour at the time), the legislator's intention was to limit the presence of Jews in the public administration of the Empire as much as possible.[98] This law also largely allows us to deduct that their presence was not anecdotical and that the military service of Jews in the army was by no means exceptional, as both facts were a cause for concern for the imperial chancellery. This was true to the point that the successor of Honorius, Valentinian III, saw fit to renew this ban, perhaps because the previous laws had not been implemented fully.[99] Indeed, the *Constitutiones Sirmondianae* contain a law promulgated at Aquileia by this emperor on July 9, 425 in which he not only reiterated the exclusion of Jews from the duties and honours of the *militia*, but he also denied them the right to *causas agere*, that is, to practice the legal profession, which they had previously enjoyed:

> [...] We also deny to the Jews, and to the pagans, the right to practice law and to serve in the State service; we do not wish people of the Christian Law to serve them, lest they substitute, because of this mastery, the venerable religion by a sect [...].[100]

This discriminatory policy would culminate shortly afterwards with the *novella* of Theodosius II in which Jews were once again barred from holding public office (including that of *defensor civitatis*) and denied every kind of honours

administration in 416 with the goal of achieving religious uniformity of the State around the Catholic creed (*CTh.*, 16.10.21). On this subject, see Chuvin ²1991, 95.

98 Linder 1987, 280; Jordán Montes 1997, 125; Braun 1998, 162. Capucine Nemo-Pekelman (2015, 31) notes: "These laws removed from the Jews certain rights such as the *ius militandi*."

99 Elsewhere, I have pointed out that "this same law leads us to think that there were enough Jews in the ranks of the army at the time for the emperor to deem their exclusion necessary. It is possible that the number of such soldiers or officers who held positions of authority in the army alarmed the Church and led to an opposite reaction that implied immediate imperial action to expel them from the army by law" (González Salinero 2000, 113).

100 *Const. Sirm.*, 6 (Linder 1987, No. 51, 307): [...] *Iudaeis quoque vel paganis causas agendi vel militandi licentiam denegamus: quibus Christianae legis nolumus servire personas, ne occasione domini sectam venerandae religionis inmutent* [...]. See Juster 1914, II, 278 note 2; Williams 1998b, 106; González Salinero 2000, 113; Stemberger 2000, 305–306; Schoenfeld 2006, 125.

and *dignitates*.[101] As Jonathan P. Roth notes, "it is difficult to imagine that these decrees do not deal with a very real presence of Jews in the Roman military."[102]

Moreover, it is worth pointing out that Jerome, in his *Commentary on Isaiah*, delighted at the memory of Jews not being allowed to serve in the army or bear arms of any sort in the early 5th century:

> [...] It is not surprising if every office of warriors was lost among the Jews, since they do not have the authority to serve as sword-bearing soldiers and to carry weapons [...].[103]

The correct interpretation of this passage has given rise to some controversy[104] as, due to the place (Bethlehem) and time (between 408 and 410 CE) it was written, it is impossible for Jerome to have been referring to the law passed by Honorius in 418 CE or to any of the other laws that were later enforced in the East. It is certainly possible that the law this Christian author had in mind was that of 404 CE, whereby the Jews of the Western part of the Empire were stripped of all civil and military honours. In fact, it would be reasonable to think that, due to the fact that most of his possible readers would have belonged to the Latin cultural sphere, he expected them to be able to easily recognize and verify the existence of the law this passage refers to. However, it is also possible that Jerome may have been describing a reality that was observable in his immediate surroundings. This would imply that, in the absence of a means of validating this Western constitution (of which there is no evidence) in the Eastern part of the Empire, there must have been a specific regulation barring Jews from joining the Eastern Roman army which, like many others, was not included in the *Codex Theodosianus*, and therefore did not survive to this day. Be that as it may, the information provided by Jerome is highly relevant, as it highlights the reaction of Imperial power against the presence of Jews in the "Christian army." The author's ultimate intention was to portray as fact the

---

101 *Nov. Th.*, III: [...] *Quam ob rem cum sententia veteri desperatis morbis nulla sit adhibenda curatio, tandem, ne ferales sectae in vitam, inmemores nostri saeculi, velut indiscreta confusione licentius evagentur, hac victura in omne aevum lege sancimus neminem Iudaeum, neminem Samaritam neutra lege constantem ad honores et dignitates accedere, nulli administrationem patere civilis obsequii, nec defensoris fungi saltem officio* [...]. See Reichardt 1978, 37–38; Williams 1998b, 106; González Salinero 2000, 113–114; Schoenfeld 2006, 125.

102 Roth 2007, 419.

103 Jerome, *In Is.*, III, 4 (ed. Gryson *et alii* 1993–1999, 223) = II, 3, 3 (ed. Adriaen 1963, 44): [...] *Nec mirum si apud Iudaeos omnis perierit dignitas bellatorum, cum militandi sub gladio et arma portandi non habeant potestatem* [...].

104 See, for example, Grayzel 1968, 97 note 11; Linder 1987, 280–281; Roth 2007, 419.

level of debasement to which the Jewish people had fallen to as a consequence of the divine punishment that inexorably befell it owing to its innumerable sins.[105] Therefore, as can be seen in patristic literature, there could be no excuse for allowing Jews to hold positions of authority or privilege within Christian society.[106] Like Jerome, many Church Fathers insisted, in their respective commentaries on *Isaiah* 2, on the allegorical fulfilment of the exclusion of Jews from the armies or their lack of troops of their own after the destruction of the Temple.[107] Theodoret of Cyrus went so far as to consider that the prophecy had been fulfilled by the 1st century, as, by that time, the Jewish people already lacked a regular army with which to face Titus and merely constituted a group of rebels.[108] Nevertheless, according to the Church, the progressive debasement of the Jews was definitively proven by the shameful loss of their status as warriors and their expulsion from the armies by the successors of Theodosius the Great. This notion was expressed thus in one such commentary, attributed to Basil of Caesarea (in Cappadocia) but which seems to have been drafted during the 5th century: "there are no longer any warriors among the Jews, nor do they feature in the lists of the army."[109] Likewise, the *Altercatio Ecclesiae et Synagogae*, written during the same century, reflects this historical reality (a direct consequence of anti-Jewish legislation then in force) when, in a dialogue between rhetorically opposed personifications of both institutions (the Church and the Synagogue), the Church pronounces the following reproach:

> [...] You pay me tribute, you are barred from command (*imperium*), you may not hold a prefecture. A Jew may not be a count (*comes*), you are forbidden entry to the Senate, you know nothing of office, you are debarred from military service, you do not come near the table of the wealthy, you have lost the order of the *clarissimus*, everything is off limits to you, and we offer you very little to be eaten, so that you might even live badly [...].[110]

---

105 On this subject see González Salinero 2003, 141–142.

106 González Salinero 2000, 106–107.

107 Fernández Ubiña 2000, 196 note 18.

108 Theodoret of Cyrus, *In Is.*, III, 1–4.

109 Pseudo-Basil, *In Is.*, 3, 2 (PG 30, 281): [...] Οὐκέτι παρὰ Ἰουδαίοις πολεμισταί. Οὐδὲ γὰρ ἐαριθμοῦνται τοῖς στρατιωτικοῖς καταλόγοις [...]. Cf. Cyril of Alexandria, *In Is.*, III, 2 (PG 70, 97) and, on legal restrictions of Jews from a broader perspective, John Chrysostom, *In Is.*, III, 2. See Juster 2014 II, 277 note 4; Lucas 1993, 65.

110 *Alt. Eccl. et Synag.*, 127–133: [...] *Tributum mihi soluis, ad imperium non accedis, habere non potes praefecturam. Iudaeum esse comitem non licet, senatum tibi introire prohibetur, praefecturam nescis, ad militiam non admitteris, mensam diuitum non adtingeris, clarissimatus ordinem perdidisti, totum tibi non licet, cui etiam ad manducandum, ut uel male uiueres,*

In spite of the above, under which circumstances of which we are unaware, Jews eventually returned to the army, this time under the Byzantine emperors. For instance, Zacharias of Mytilene (ca. 465-after 536 CE)—also known as Zacharias Scholasticus or Zacharias Rhetor—tells us of their service in the armies of the Eastern Roman Emperor Marcian (450–457)[111] and the *Chronicle* of Pseudo-Joshua the Stylite writes of their active participation in the local military forces of the city of Tella against the Persian attacks of the years 502–503 CE.[112] Nevertheless, Justinian and Justin decided to return to the spirit of earlier Christian legislation, banning both Jews and Samaritans from the army:

> [...] Indeed we order that those who are heretics, and above all the pagans, Jews, Samaritans, and those similar to them, if they take part in any of all those we have already recalled, having obtained an honour, inscribed in the advocates list, taken an office or put on an official belt, they shall be thrown out on the spot from participating in these. For we want all the above-mentioned to be purged from association with such as these now and for ever, not only in this glorious city, but in practically every province and every place. [...][113]

---

paucula condonamus. [...]. See Castritius 2003, 7. As Arthur Lukyn Williams (1935, 337) has pointed out, the *Altercatio* seems to be some sort of pseudo-legal and apparently provincial text, judging by the language used. Its date of composition is uncertain, but the evidence points to it having been drafted between 438 and 476 (Blumenkranz 1963, 40–41). González Salinero 2019b.

111 Zacharias of Mytilene, *Hist. eccl.*, III, 5–6 (CSCO, 87, 109–110).

112 *The Chronicle of Pseudo-Joshua the Stylite*, 58. See Juster 1914, II, 279; Rabello 1987–1988, I, 105, note 55. In fact, this would not be the only time in which Jews joined local armies in moments of difficulty, when their collaboration was requested to defend cities. With regards to the Hellenistic period, we know through Flavius Josephus that this was a very common occurrence (Juster 1914, II, 278; Askowith 1915, 203). In the West, for instance, they participated in the local militia during the defense of Arles in 508 (Cyprian of Toulon, *Vit. Caes.*, I, 3, 21–22: PL 67, 1011–1012 = *MHG, script. rer. mer.*, III, 468) and during the defense of Naples against Byzantine troops in 536 (Procopius of Caesarea, *De bell. goth.*, III, 41–45). On this subject, see Juster 1914, II, 278–279; Blumenkranz 1960, 35.

113 *Cod. Iust.*, 1.5.12 (527 CE): [...] Ἀλλὰ καὶ τοὺς ὄντας μὲν αἱρετικοὺς καὶ ἔτι πρὸ τούτων Ἕλληνας ἢ Ἰουδαίους ἢ Σαμαρείτας καὶ τοὺς τούτοις ὁμοίους, μετασχόντας δέ τινος ἤδη τούτων, ὥνπερ ἐμνήσθημεν, καὶ τυχόντας ἀξιώματος ἢ τοῦ καταλόγου τῶν δίκας ἀγοπευόντων ἢ στρατείαν περιβεβλημένους ἢ ζώνην ὁποιανοῦν ἐκβληθῆναι τῆς πρὸς ταῦτα μετουσίας παραχρῆμα παρακελευόμεθα. Καθαρεύειν γὰρ τὰ ῥηθέντα πάντα τῆς τών τοιούτων νῦν τε καὶ διά παντὸς βουλόμεθα κοινωνίας, οὐκ ἐπὶ μόνης ταύτης τῆς ἐνδόξου πόλεως, ἀλλ' ἐπὶ πάσης ὅλως ἐπαρχίας καὶ τόπου παντός. [...]. Cf. *Cod. Iust.*, 1.5.18 (529 CE). John Malalas, *Hist.*, XI (PG 97, 65 ff.) also refers to the official exclusion of Jews from the army. See Juster 1914, II, 278 note 2; Rabello 1987–1988, I, 241 note 2; 377 note 3; II, 679, 720 and 726; de Lange 2005, 414 and 421; Kraemer 2020, 295.

We are unable to judge the effectiveness of these constitutions in the Eastern Roman Empire, but the fact remains that, by the 6th century, the Jews still maintained their reputation as good fighters[114] and, in fact, they continued to serve in special units of the various armies until the 11th century at the earliest.[115]

114 Kraemer 2020, 325, 353.

115 Blumenkranz 1960, 34–36; Rabello 1987–1988, I, 204, note 4.

Conclusion

> […] very probably he had long since come to the sage and sensible conclusion that a man's religion is one thing, and this practical world quite another […].
>
> HERMAN MELVILLE, *Moby Dick*, ch. XVI

∵

The military service of Jews in the Hellenistic kingdoms, the evidence for which has been repeatedly highlighted in research, has been widely acknowledged. Conversely, however, the possible collaboration of Jews with Roman power, which has been obvious from the time of Herod at the very least, and their sometimes forced, sometimes voluntary presence in the Roman army have been viewed with a certain degree of scepticism.[1] The limited number of surviving inscriptions and the dismissal or even insufficient or improper analysis of written sources that specifically cover the military service of Jews in the Roman army have long been an obstacle to a qualitative advance in our knowledge of this phenomenon, which has been ignored for so long. The challenges of identifying Jewish individuals with Greek or Latin names who are not identified as such in inscriptions, even through symbols (menorah, lulav, shofar, palm), common epigraphic formulas in Jewish settings (ἐνθάδε κεῖται, *Deo Aeterno*) or unmistakably Jewish ones (ἐν εἰρένη ἡ κοίμησις αὐτοῦ, *sit pax in requie eius*, 'In peace his/her/your sleep', *in pace, cum pace, dormitio eius cum iustis* …), have likewise discouraged certain researchers with a tendency to avoid contextual analysis. Many such inscriptions have been found in Jewish catacombs, such as those of Rome or Beth She'arim. There are even cases that feature typically gentile funerary formulae, such as *D*(*is*) *M*(*anibus*)[2] or that have been found in catacombs decorated with pagan motifs, such as those of Vigna Randanini.[3] Nevertheless, we must keep in mind that, as Leonard Rutgers notes, "Jews availed themselves of the trappings of Greco-Roman society, but they did so in order to express an identity that was patently and unmistakably Jewish."[4]

1 Schoenfeld 2006, 117, Roth 2007, 420.

2 *JIWE 1*, 285–288; *JIWE 2*, 489–495; *IJO I*, 7–8, 15, 21, 390.

3 Laurenzi 2011, 53–57.

4 Rutgers 1998a, 39.

DOI:10.1163/9789004507258_007

On the other hand, many researchers tend to rely exclusively on rabbinic literature when it comes to reconstructing a universal image of "normal" Jewish practices in the Roman world. This has hindered the study of other expressions or currents of ancient Judaism which were particularly vibrant in the Greco-Roman diaspora, and which should not necessarily be deemed to be marginal or "heterodox" manifestations that distorted "true" Judaism.[5]

We cannot say that Jews joined the Roman army in massive numbers, but their presence was by no means insignificant.[6] We could accept that Jews who chose to pursue a career in the military were not particularly pious, and that they were forced to relax significantly their observance of religious principles unique to their Jewish tradition, but that did not prevent them from considering themselves Jewish. In fact, there is not a shred of evidence to justify describing them as Jewish "renegades". There is no evidence that joining the Roman military world necessarily implied betraying their religion, even during the most bitter moments of the Jewish revolts, as there were always two trends within Judaism, one of which was closer to the Roman order, deemed to be established on Earth by divine design, and the other of which was in open confrontation with the invading power as an enemy of the Jewish people. For this reason, rather than a war of liberation or independence struggle, Jonathan P. Roth prefers to describe the first Jewish war "as a civil war."[7]

The diversity of beliefs and practices that existed in the Jewish cultural world and which were not entirely stamped out by the so-called "normative" Judaism that was gradually imposed by the rabbinic movement, stands in the way of universal statements on the definition of Jewish religion and identity in the Greco-Roman world. There were customs, rites and precepts, such as circumcision, Sabbath rest or the avoidance of certain foods that could legitimately be held as essential traits of Judaism. However, whereas circumcision was an irreversible "sign", the other formal aspects of Jewish ritual practice were often subject to circumstances and, therefore, to exceptions, variations and even differing interpretations.[8] Synagogal communities, both in Eretz Israel and in the Diaspora, featured a significant variety of conceptions of the Jewish faith and

5 Schoenfeld 2006, 117. Cf. Friedheim 2015; Rutgers 2016. Furthermore, it does not appear that the rabbinical movement spread rapidly, either in Israel or the Diaspora, nor that it was dominant everywhere. See now Costa 2019 and Laham-Cohen 2021.

6 Oppenheimer 2005, 187. There may have been cases of Jews who avoided compulsory drafts through the use of a substitute (*vicarius*).

7 Roth 2019, 93.

8 For instance, Tal Ilan (2009) proved that, as is shown by *Pseudo-Philo* in Latin, the study of Torah in the Jewish community in Rome was strongly influenced by certain rabbinic traditions and teachings that were not interested in the Law as much as in displaying a

worship during the Greco-Roman era: "pompous versus ascetic, large versus tiny, and varying degrees of foreign influence absorbed into the overall architecture, or inside decoration, which, in many cases, included wall-paintings and mosaics—which many may have considered as a sin contradicting the second commandment—and in some cases even a clear pagan influence, in the form of the zodiac."[9] No behaviour deemed 'unusual' by the standards of the religious norms of rabbinic Judaism—which, in turn, contained significant doctrinal diversity—[10] necessarily implied a renunciation of a Jewish cultural and religious heritage. Despite the fact that the observance of certain Jewish customs and religious practices was, in theory, incompatible with military life, the evidence of the active participation of Jews in the Roman army requires us to consider the existence of a broad scope of possibilities. These range from the assumption that Jewish soldiers may have, at certain times, benefitted from a privileged legal status that exempted them from some formalities, such as the military oath, to the notion that they themselves may have renounced the strict observance of the precepts of their religion which, like the Sabbath rest, prevented them from carrying out military activities at certain times, without renouncing their Jewishness.

The support they provided to Julius Caesar during his civil war against Pompey and their loyalty towards Roman power from the reign of Herod onwards allowed Jews to maintain and even strengthen their 'vertical alliance.' As Louis H. Feldman notes, "the fact that no fewer than eight cities in Asia Minor were pressured by the Romans to stop their harassment of the Jews indicates both how deeply these privileges were resented locally and how strongly the central Roman government felt about protecting the Jews."[11] These privileges doubtless included an exemption from military service in Roman auxiliary forces. This favourable treatment of Jews was undoubtedly eroded during the Jewish Wars, but it never disappeared entirely. Once again, to quote Louis H. Feldman's

deep acquaintance with extra-biblical rabbinic interpretations, which could be defined as *midrash*—exposition.

9 Olshanetsky 2015, 21. On the diversity of Jewish thought on artistic representations in ancient Judaism, see the studies included in Pearce 2013.

10 According to Catherine Hezser (2016, 205), "those who affiliated with a particular rabbi were not necessarily adherents of all rabbis or "rabbinic Judaism" as such. [...] The collective, vertically-oriented Temple cult had been replaced by a horizontal network based on one's personal affiliation with a specific representative of the holy whose relationship to other representatives was ambiguous and competitive."

11 Feldman 1993, 423. The occasional exceptions to this policy are blamed by Philo on a corrupt minister—such as the ambitious praetorian prefect Sejanus—or on an insane emperor (see *Legatio ad Gaium*, 159–165, 184–189, 338–348).

insightful comments, "even after long and bloody revolt of 66–74 the Romans refused to remove the special privileges of the Jews in such important cities as Antioch and Alexandria, despite the insistent request of the non-Jewish population. Again, even after the great Diaspora revolt of 115–117, the Romans did not change the fundamentals of their policy. When, to be sure, after the Bar Kochba rebellion of 132–135 the Emperor Hadrian forbade many Jewish observances, these edicts, too, were soon alleviated by his successor, Antoninus Pius [...]."[12]

There is no evidence that Augustus included a *general* exemption from military service among the privileges granted to Jews in the Roman Empire. Jews living in the cities of the Levant may have invoked the legal precedent to avoid local recruitment drives, but this was not the case in the West. Under Tiberius, they were forced to perform military service in Sardinia and were not allowed to submit any legal objections to this decision which, according to pagan historians, was punitive in nature. In any event, the exemption from military service never precluded voluntary enlistment in order to reap the benefits of Roman citizenship. There are some cases of Jews who obtained it through *honesta missio*, according to their military diplomas. In fact, the presence of Jews can be seen in the Roman troops that intervened in a number of conflicts after the Bar Kochba rebellion. According to Cassius Dio and the *Historia Augusta*, they served under Avidius Cassius and Pescenius Niger, respectively, and it seems that their numbers began to swell during the Severan period until they reached considerable numbers in the 4th century CE. The fact that they had infiltrated the *schola agentum in rebus* and even become officers in the *militia armata* was cause for alarm for the Church hierarchy which, inflamed by an increasingly vehement anti-Jewish ideology, could not tolerate the notion of Jewish officers commanding Christian soldiers, or the "deicide people" being able to defend the Christian people. Giving in to the influence of the Church and ignoring the legal tradition according to which Judaism was a *licita religio*,[13] the Empire began to enact a series of laws through which Jews were excluded from all positions and honours acquired within the Roman army.

Despite the exemption from military service granted to Jews at a given moment by the Roman authorities in thanks for their loyalty, they were not prevented from voluntarily joining the army and, therefore, reaping the benefits of *honesta missio*. However, their expulsion from the *militia armata* during

12 Feldman 1993, 424. Cf. Burr 2020, 125–126 (orig. 1955, 81).

13 *CTh.*, 16.8.9 (29 September 393): *Iudaeorum sectam nulla lege prohibitam satis constat* ("It is sufficiently established that the sect of the Jews is prohibited by no law.") See Linder 1987, 189–191; González Salinero 2000, 78–79.

the Christian Empire contributed significantly to their social exclusion in a new order in which they had no place. Paradoxical as it may seem, the presence of Jewish soldiers in the Roman army started from the will to renounce a privilege—that of exemption—and ended with a legal obstacle based on their radical exclusion.

APPENDIX 1

Violence and the Use of Arms on Sabbath

The main obstacle to the service of Jews in the Roman arms was, at least officially, the incompatibility of their religious practices with military life.[1] As we saw earlier, P. Cornelius Dolabella expressly pointed to this reason as an apparent justification for his decree.[2] In theory, both dietary laws (*kashrut*) and keeping the Sabbath rest would prevent Jews from becoming fully integrated in military life.[3] In fact, strict observance of the Sabbath rest would even prevent soldiers serving in purely Jewish units or participating in the defence of a Jewish city from bearing arms on that day.[4] And yet, over the course of centuries, historical circumstances forced Jews to adapt the relationship between Sabbath and war to a broad and ever-changing set of exceptional cases.[5]

Although they were aware that this precept was due to a purely religious scruple,[6] pagan authors such as Plutarch deemed the refusal to bear arms on Sabbath to be an obsessive and absurd behaviour, which could only lead to disaster in desperate circumstances, such as the defence of a besieged city:

> But the Jews, because it was the Sabbath day, sat in their places immovable, while the enemy were planting ladders against the walls and capturing the defences, and they did not get up, but remained there, fast bound in the toils of superstition as in one great net.[7]

---

1 Josephus, *Ant. Iud.*, XIII, 251–252; XIV, 226; XVIII, 84; *Contr. Ap.*, I, 209. See Schürer 1973, I, 363.

2 Josephus, *Ant. Iud.*, XIV, 226–227. See Nodet 1997, 89; Batsch 2005, 251.

3 Josephus, *Ant. Iud.*, XIV, 226; *Mishnah*, *Shab.*, VI, 4 (although it is possible to make amends through a sacrifice). See Schürer 1979, II, 475 note 62; Wilde 1949, 30; Herr 1960–1961, 354–356. Shedding the blood of the enemy on Sabbath, even in legitimate self-defense, was always tainted by "impurity" (Batsch 2005, 41–56 and 87–116; Le Bohec 2015, 186). Therefore, according to the *Sibylline Oracles* (III, 727–730), the establishment of the Messianic Age was characterised by the absolute dominion of weapons. See Nikiprowetzky 2000, 5.

4 On the prohibition to bear arms on Sabbath, Mishnah, *Sabbath*, VI, 3–4; Tosefta, *'Erubin*, III, 5–6; JT, *'Erubin*, IV, 3 (21d); BT, *'Erubin*, 45a; Josephus, *Vita*, 161. *See* Herr 1960–1961, 354–356; Nodet 1997, 84.

5 Nikiprowetzky 2000, 6 ff.

6 In the late 1st century CE, Frontinus understood the nature of this precept as it commented that it was "a day on which it is sinful (*nefas*) for them to do any business, and so defeated them" (*Strategemata*, II, 1, 17). See Feldman 1993, 161; McKay 1994, 109.

7 Plutarch, *De superstitione*, 8 (169C, 10): ἀλλ' Ἰουδαῖοι σαββάτων ὄντων ἐν ἀγνάμπτοις καθεζόμενοι, τῶν πολεμίων κλίμακας προστιθέντων καὶ τὰ τείχη καταλαμβανόντων, οὐκ ἀνέστησαν ἀλλ' ἔμειναν ὥσπερ ἐν σαγήνῃ μιᾷ τῇ δεισιδαιμονίᾳ συνδεδεμένοι. On the astonishment this Jewish custom

 | DOI:10.1163/9789004507258_008

The historian Agatharchides of Cnidus spoke of the Jewish refusal to defend Jerusalem on Sabbath from the army of Ptolemy I Soter (ca. 302/301 BCE) in a text reproduced by Josephus:

> The people known as Jews, who inhabit the most strongly fortified of cities, called by the natives Jerusalem, have a custom of abstaining from work every seventh day; on those occasions they neither bear arms nor take any agricultural operations in hand, nor engage in any other form of public service, but pray with outstretched hands in the temples until the evening. Consequently, because the inhabitants, instead of protecting their city, persevered in their folly, Ptolemy, son of Lagus, was allowed to enter with his army; the country was thus given over to a cruel master, and the defect of a practice enjoined by law was exposed.[8]

In the Torah, the Sabbath rest stands out as one of the essential commandments of Judaism, which had become one of the defining traits of Jewish identity since ancient times.[9] Nevertheless, as Christophe Batsch points out, the Hebrew Bible does not establish any specific rules on war and the use of arms during that day of the week. The activities forbidden in the Ten Commandments (*Ex* 20:8–11, 31:14 and *Dt* 5:12–15) are essentially peaceful, and there is no mention of renouncing struggle in *Jubilees* 50:8–13, the first list in Jewish literature of the activities banned on Sabbath.[10] Thus, any

---

caused among Classical writers, see Michael 1924; Feldman 1993, 160; and Pucci ben Zeev 2016, 124. Nevertheless, even though the interruption of all daily activity for a certain period of time for religious reasons is usually associated to the Jews, such behaviour is not exclusive to them. For instance, in the Greek world, the *Hieromenia* refers to the sacred time of a festival and signals of restrictions on regular daily activity. We also know that military campaigns were interrupted by important festivals and that sometimes Greeks were attacked just because they were celebrating a festival and so were unprepared for fighting (Herodotus, VI, 87; Thucydides, III, 3, 3; Xenophon, V, 2, 25–26). See Parker 1983, 157; Harrington 2001, 115.

8 Josephus, *Contr. Ap.*, I, 209–210: "οἱ καλούμενοι Ἰουδαῖοι πόλιν οἰκοῦντες ὀχυρωτάτην πασῶν, ἣν καλεῖν Ἱεροσόλυμα συμβαίνει τοὺς ἐγχωρίους, ἀργεῖν εἰθισμένοι δι' ἑβδόμης ἡμέρας καὶ μήτε τὰ ὅπλα βαστάζειν ἐν τοῖς εἰρημένοις χρόνοις μήτε γεωργίας ἅπτεσθαι μήτε ἄλλης ἐπιμελεῖσθαι λειτουργίας μηδεμιᾶς, ἀλλ' ἐν τοῖς ἱεροῖς ἐκτετακότες τὰς χεῖρας εὔχεσθαι μέχρι τῆς ἑσπέρας, εἰσιόντος εἰς τὴν πόλιν Πτολεμαίου τοῦ Λάγου μετὰ τῆς δυνάμεως καὶ τῶν ἀνθρώπων ἀντὶ τοῦ φυλάττειν τὴν πόλιν διατηρούντων τὴν ἄνοιαν, ἡ μὲν πατρὶς εἰλήφει δεσπότην πικρόν, ὁ δὲ νόμος ἐξηλέγχθη φαῦλον ἔχων ἐθισμόν […]." Cf. Josephus, *Ant. Iud.*, XII, 5–6. See Johns 163, 483; Feldman 1993, 159; Batsch 2005, 247, 273; Bar-Kochva 2010, 291–296.

9 In the words of Tessa Rajak (1985, 17), "the Sabbath was central to the identity of Jews living among Gentiles." John Barclay (1996, 443–444) also extends the central bond of Jewish identity to circumcision, separation at meals and rejection of alien cultic activity.

10 Doering 1997, 182; Nodet 1997, 63; Batsch 2005, 255. According to Julia Rhyder (2019), the Holiness legislation creatively blends two aspects of the Sabbath that were not always con-

considerations on the rejection of armed struggle on Saturday must necessarily stem from an interpretation of Biblical texts,[11] in view of the ban on long trips[12] or any other activity that disturbed the Sabbath rest, and which necessarily entailed a ban on fighting.[13] It seems that, during the time period discussed by the passage from Agatharchides of Cnidus quoted by Josephus (late 4th century BCE) the ban on fighting under all circumstances on Sabbath already had widespread acceptance in the Jewish tradition.[14] Nevertheless, we are unaware of its origins, which do not seem to go back to the pre-exilic period, as Alger Johns believed when he argued that the final conquest of Jerusalem by Nebuchadnezzar on 29 July 587 BCE took place on a Saturday based on the inaccurate *Chronicles of Chaldaean Kings*.[15] According to André Lemaire, the establishment of Saturday as the seventh day of the week given over to rest was not instituted in Judaism until the Return to Zion, under Persian rule, owing to the reorganization of the priesthood promoted by Nehemiah and Ezra.[16] In fact, we know nothing of Jewish observance of the Sabbath when faced with a foreign menace until the conquest of Jerusalem by the armies of Ptolemy I, and not even this episode proves the existence of a consensus among period Jews in this regard. Indeed, the version of the events given by Josephus in his *Antiquitates Iudaicae* casts doubt on the matter:

> And this king seized Jerusalem by resorting to cunning and deceit. For he entered the city on the Sabbath as if to sacrifice, and, as the Jews did not oppose him—for they did not suspect any hostile act—and, because of their lack of suspicion and the nature of the day, were enjoying idleness and ease, he became master of the city without difficulty and ruled it harshly.[17]

---

nected: (1) the idea, already present in the Decalogue and Gen 2:2–3, that the Sabbath is a time of cessation held every seventh day; and (2) the more traditional associations of the Sabbath with sacrificial rites at the shrine.

11 Batsch 2005, 248.

12 See Hezser 2011, 51–52 and 65–67.

13 Josephus, *Ant. Iud.*, XIV, 226. See Batsch 2005, 260, 272.

14 Beyond the cases that arose at the time of the Maccabees, Josephus (*Ant. Iud.*, XVIII, 319) mentions another episode of this nature during the rule of Ptolemy IX Lathyrus (116–80 BCE). Cf. *Ant. Iud.*, XVIII, 354. See Nikiprowetzky 2000, 8.

15 John 1963, 483–384 and 486. See also Goodman and Holladay 1986.

16 Lemaire 1973. As Robert Goldenberg (1979, 431 note 68) already pointed out, "the Bible gives no hint of a pre-exilic ban on warfare on the Sabbath." Cf. Batsch 2005, 248.

17 Josephus, *Ant. Iud.*, XII, 4: κατέσχε δὲ οὗτος καὶ τὰ Ἱεροσόλυμα δόλῳ καὶ ἀπάτῃ χρησάμενος· ἐλθὼν γὰρ σαββάτοις εἰς τὴν πόλιν ὡς θύσων, μήτε τῶν Ἰουδαίων αὐτὸν ἀμυνομένων, οὐδὲν γὰρ ὑπενόουν πολέμιον, καὶ διὰ τὸ ἀνύποπτον καὶ τὴν ἡμέραν ἐν ἀργίᾳ καὶ ῥαθυμίᾳ τυγχανόντων, ἀπόνως ἐγκρατὴς γίγνεται τῆς πόλεως καὶ πικρῶς ἦρχεν αὐτῆς.

Thus, Josephus' version of the events described by Agatharchides does not allow us to say conclusively that the Jews would not have fought on Sabbath had they discovered that the Ptolemaic king's pious intentions were false. The Jewish author insists that Ptolemy I took the city by means of a ruse, not because its inhabitants declined to defend it due to the Sabbath rest. In fact, it even conveys the impression that, had they remained alert, the Jews would have defended themselves through the use of arms, as Josephus makes no mention of the city walls and gates, which were obviously open to allow the king to enter and offer the sacrifice he promised.[18]

The debate on the observance of the Sabbath rest under any circumstances and the possible legitimacy of the use of arms on that sacred day for defensive reasons reached its most crucial juncture with the Maccabean revolt of 167 BCE.[19] Early in the conflict, the leaders of the Seleucid armies followed a strategy of attacking the defenceless Jews on Saturdays.[20] According to the author of the first book of *Maccabees*, some of them (the so-called Ḥasidim, Hasidaeans) decided to flee to the caves in the desert to avoid persecution and observe the rules of their religion. Nevertheless, when they were discovered by the army of Antiochus Epiphanes, they were cruelly butchered:

> Without more ado the attack was launched; but the Israelites did nothing in reply; they neither hurled stones, nor barricaded their caves. 'Let us all meet death with a clear conscience,' they said; 'we call heaven and earth to witness it is contrary to all justice that you are making way with us.' So on the sabbath they were attacked and massacred, men, women, and children, up to a thousand in all, along with their livestock.[21]

---

18 Nodet 1997, 65–67.

19 Goldenberg 1979, 432; McKay 1994, 47; Batsch 2005, 306.

20 Sekunda 1994, 17.

21 *1Mac* 2:35–38: καὶ ἐτάχυναν ἐπ' αὐτοὺς πόλεμον. καὶ οὐκ ἀπεκρίθησαν αὐτοῖς οὐδὲ λίθον ἐνετίναξαν αὐτοῖς οὐδὲ ἐνέφραξαν τοὺς κρύφους λέγοντες Ἀποθάνωμεν πάντες ἐν τῇ ἁπλότητι ἡμῶν· μαρτυρεῖ ἐφ' ἡμᾶς ὁ οὐρανὸς καὶ ἡ γῆ ὅτι ἀκρίτως ἀπόλλυτε ἡμᾶς. καὶ ἀνέστησαν ἐπ' αὐτοὺς ἐν πολέμῳ τοῖς σάββασιν, καὶ ἀπέθανον αὐτοὶ καὶ αἱ γυναῖκες αὐτῶν καὶ τὰ τέκνα αὐτῶν καὶ τὰ κτήνη αὐτῶν ἕως χιλίων ψυχῶν ἀνθρώπων. According to the summarized account in *2Mac* 6:11: "Other Jews, who had assembled secretly in nearby caves to observe the sabbath, were denounced to Philip and, since out of regard for the sanctity of the day thy had scruples about defending themselves, they were burnt alive" (ἕτεροι δὲ πλησίον συνδραμόντες εἰς τὰ σπήλαια λεληθότως ἄγειν τὴν ἑβδομάδα μηνυθέντες τῷ Φιλίππῳ συνεφλογίσθησαν διὰ τὸ εὐλαβῶς ἔχειν βοηθῆσαι ἑαυτοῖς κατὰ τὴν δόξαν τῆς σεμνοτάτης ἡμέρας.) See Bar-Kochva 1989, 481–482; Henten and Avemarie 2006, 34–38.

This strategy[22] seems to have served the Seleucids well until Mattathias, one of the first leaders of the Maccabean movement, promoted the reform of the 'Halakhah'.[23] According to Josephus, Mattathias was able to convince the Jews after the massacre to fight the enemy even on Sabbath, based on the new legal precept (τὸ νόμιμον) he established:[24]

> [...] And he [Mattathias] instructed them to fight even on the Sabbath, saying that if for the sake of observing the law they failed to do so, they would be their own enemies, for their foes would attack them on that day, and unless they resisted, nothing would prevent them from all perishing without striking a blow.[25]

Faced with the peril of an enemy attack on Saturday and the lack of clear precedents regulating proper behaviour according to the law, the rebels led by Mattathias ultimately decided to respond to aggression by the use of arms, despite the fact that this resolve implied a blatant breach of the Sabbath rest.[26] Even though Josephus states that "to this day we continue the practice of fighting even on the Sabbath whenever it becomes necessary,"[27] and there were, indeed, occasions on which the Sabbath rest was not observed,[28] there was also no shortage of cases in which Jews, ignoring Mattathias' reform, preferred defeat to breaking the Sabbath.[29] Cassius Dio states that Jerusalem

---

22 Le Bohec 2015, 179–180.

23 Schürer 173, I, 143; Schürer 1979, II, 474; Goldenberg 1979, 431–432; Kubo 1982, 61–65; Bringmann 1983, 138–139; Bar-Kochva 1989, 474–493; McKay 1994, 45–50; Nodet 1997, 63–64; Nikiprowetzky 2000, 2–7; Batsch 2005, 256–270; Seeman 2013, 59; Tonstad 2016.

24 Josephus, *Ant. Iud.*, XII, 276.

25 Josephus, *Ant. Iud.*, XII, 276: [...] ὁ δὲ καὶ σαββάτοις αὐτοὺς ἐδίδαξε μάχεσθαι λέγων, ὡς εἰ μὴ ποιήσουσι τοῦτο φυλαττόμενοι τὸ νόμιμον, αὐτοῖς ἔσονται πολέμιοι, τῶν μὲν ἐχθρῶν κατ' ἐκείνην τὴν ἡμέραν αὐτοῖς προσβαλλόντων, αὐτῶν δ' οὐκ ἀμυνομένων, κωλύσειν τε μηδὲν οὕτως ἀμαχητὶ πάντας ἀπολέσθαι [...].

26 *1Mac* 2:32–34, 1; Josephus, *Ant. Iud.*, XII, 268 ff.

27 Josephus, *Ant. Iud.*, XII, 277: [...] καὶ ἄχρι δεῦρο μένει παρ' ἡμῖν τὸ καὶ σαββάτοις, εἴ ποτε δεήσειεν, μάχεσθαι. Cf. Josephus, *Bell. Iud.*, II, 517. See Feldman 1993, 160–161; Batsch 2005, 276. Nevertheless, Josephus does not specify if these were merely defensive actions, as if he were attempting to conceal or justify his participation in necessary military actions that may have taken place on a Saturday (Nodet 1997, 64).

28 Josephus, *Ant. Iud.*, XIII, 10–13; XIII, 252–253; XIII, 337; XIV, 63–64; XVIII, 318–324; *Bell. Iud.*, II, 456–457; II, 517; II, 634. However, we know that, during the Hasmonean period, both John Hyrcanus and his successors often hired mercenaries (who were not bound by the Sabbath rest) to fight offensive wars. See McKay 1994, 45–50; Nodet 2019, 63.

29 *2Mac* 5:25–26; 6:11; 8:25–27; 9:34; 9:43; 10:33–38; 12:31–32; 12:38; Josephus, *Bell. Iud.*, VII, 52; *Ant.*, XIV, 63–64. See McKay 1994, 78; Batsch 2005, 260–262; Horbury 2014, 158–159; Mor 2016, 428.

was taken by the Romans on a Saturday up to three times because its defenders did not fight to save it: by Pompey (*Hist.*, XXXVII, 16, 1–4), Mark Antony (*Hist.*, XLIX, 22, 3–5) and Vespasian (*Hist.*, LXV, 7, 2).[30] In the first episode involving Pompey in the year 63 BCE,[31] Cassius Dio admits that, had the Jews defended the wall that surrounded the Temple of Jerusalem on all days alike, the Roman general could not have got possession of it.[32] However, he adds that "they made an exception of what are called the days of Saturn, and by doing no work at all on those days afforded the Romans an opportunity in this interval to batter down the wall."[33] Josephus provides more detailed information that allows for another interpretation of the episode:

> Pompey himself was on the north side, engaged in banking up the fosse and the whole of the ravine with materials collected by the troops. The tremendous depth to be filled, and the impediments of every sort to which the work was exposed by the Jews above, rendered this a difficult task. Indeed, the labours of the Romans would have been endless, had not Pompey taken advantage of the seventh day of the week, on which the Jews, from religious scruples, refrain from all manual work, and then proceeded to raise the earthworks, while forbidding his troops to engage in hostilities; for on the sabbaths the Jews fight only in self-defence.[34]

We can infer from this text that Pompey was well aware of the Jewish custom regarding the Sabbath rest, as he was aware that the only way to ensure that his troops would not be harassed by the defenders was to abstain from any attacks that might trigger a violent reaction. He knew the Jews could not prevent him from building siege works on Sabbath because they were only allowed to bear arms to defend themselves, and his

30 Cf. Tacitus, *Hist.*, V, 4. See Phang 2012, 215.

31 Cameron 2018.

32 Nikiprowetzky 2000, 8.

33 Cassius Dio, *Hist.*, XXXVII, 16, 2: [...] νῦν δὲ τὰς τοῦ Κρόνου δὴ ὠνομασμένας διαλείποντες, καὶ οὐδὲν τὸ παράπαν ἐν αὐταῖς δρῶντες, παρέδωκαν τοῖς Ῥωμαίοις καιρὸν ἐν τῷ διακένῳ τούτῳ τὸ τεῖχος διασεῖσαι. Accoding to Strabo (*Geogr.*, XVI, 2, 40), "Pompey seized the city, it is said, after watching for the day of fasting, when the Judaeans were abstaining from all work [...]" (κατελάβετο δ', ὥς φασι, τηρήσας τὴν τῆς νηστείας ἡμέραν, ἡνίκα ἀπείχοντο οἱ Ἰουδαῖοι παντὸς ἔργου [...]). See McCasland 1943, 65–66; Goldenberg 1979, 431; Feldman 1993, 161.

34 Josephus, *Bell. Iud.*, I, 146: κἂν ἀτέλεστος ἔμεινεν τοῖς Ῥωμαίοις ὁ πόνος, εἰ μὴ τὰς ἑβδομάδας ἐπιτηρῶν ὁ Πομπήιος, ἐν αἷς παντὸς ἔργου διὰ τὴν θρησκείαν χεῖρας ἀπίσχουσιν Ἰουδαῖοι, τὸ χῶμα ὕψου τῆς κατὰ χεῖρα συμβολῆς εἴργων τοὺς στρατιώτας· ὑπὲρ μόνου γὰρ τοῦ σώματος ἀμύνονται τοῖς σαββάτοις. ἤδη δὲ ἀναπεπληρωμένης τῆς φάραγγος πύργους ὑψηλοὺς ἐπιστήσας τῷ χώματι καὶ προσαγαγὼν τὰς ἐκ Τύρου κομισθείσας μηχανὰς ἐπειρᾶτο τοῦ τείχους.

troops were safe as long as they did not directly provoke the Jews into responding. Thus, we cannot state that the Jews had already forgotten Mattathias' reforms by then.[35] Pompey's ruse allowed him to circumvent the only case that allowed the besieged to defend their city, ultimately leading to its downfall.[36]

Christophe Batsch argued that it is a mistake to apply an analysis based on an opposition between "defensive" and "offensive" wars to the various opinions on the use of arms on Sabbath, assuming that the opposition between the above notions only emerged in the Modern era after it was formulated by Hugo Grotius (1583–1645) in *De juris belli ac pacis*.[37] Nevertheless, as emerges from the text by Josephus quoted above, the distinction between defensive and offensive actions ultimately determined the decision to fight or to keep the Sabbath whatever the consequences. It is true that Philo of Alexandria (ca. 15 BCE–ca. 50 CE) deemed this distinction to be irrelevant, as they were opposed to the use of war as an excuse to break the Sabbath rest under all circumstances.[38] Nevertheless, a presumably historical episode described by Philo[39] reveals the existence of a different *Halakhah*[40] at the time, which was, moreover, probably known to the Romans. Philo describes an attempt by a high-ranking civil servant in Egypt to convince the Jews of Alexandria to renounce the Sabbath should extreme circumstances arise. The Roman civil servant listed war as one of the force majeure causes that would allow the Sabbath rest to be broken: "Suppose," he said, "there was a sudden inroad of the enemy".[41] Although we are unaware of the speaker's intentions when he attempted to convince the Jews to break the Sabbath if required (perhaps to have them set aside their scruples against waging war against alleged enemies?), the fact is that, in a way, this episode illustrates the diversity of opinions on the subject that had existed in Judaism since the 2nd century BCE at the latest.[42] In fact, we know

35 Nodet 1997, 72; Instone-Brewer 2011, 18 note 14.

36 See Canfora 2021, 33 and 229.

37 Batsch 2005, 270. Furthermore, this author wonders if the legitimacy of defensive struggle on Sabbath might be a reflection of the *Halakhah* of Josephus and his group applied to the 1st century BCE (Batsch 2005, 283).

38 Batsch 2005, 301. Cf. Weiss 1991.

39 Philo, *De somniis*, II, 123–129. The author himself states that he witnessed this event: χθὲς δ' οὐ πρῴην ἄνδρα τινὰ οἶδα τῶν ἡγεμονικῶν (*De somniis*, II, 123).

40 According to Jacob Neusner (2016, 289), "The halakhah in its way makes exactly the same statement about the same matters that the aggadah does in its categories and terms. But the aggadah speaks in large and general terms to the world at large, while the halakhah uses small and particular rules to speak to the everyday concerns of ordinary Israelites; the aggadah addresses exteriorities, the halakhah, interiorities, of Israel in relationship with God."

41 Philo, *De somniis*, II, 125: εἰ πολεμίων ἔφοδος αἰφνίδιον γένοιτο. See Batsch 2005, 298–299.

42 Doering 1997, 273. As Christophe Batsch (2005, 269) stated, "cette diversité de décisions exclut l'hypothèse qu'on soit parvenu à une position commune à tout le judaïsme, avant

that the *Halakhah* that prioritised saving one's life over keeping the Sabbath had taken on tremendous force as early as the first half of the 1st century CE, when the episode described by Philo must have taken place.[43]

Even though breaking the Sabbath was only justified in a defensive war,[44] Josephus also provides valuable information on other interpretations according to which Jews were allowed to carry out both defensive and offensive military operations on the Sabbath.[45] In fact, in some passages he cannot conceal the attacks initiated by Jews on their holy day:

> The Jews, seeing the war now approaching the capital, abandoned the feast and rushed to arms; and, with great confidence in their numbers, sprang in disorder and with loud cries into the fray, with no thought for the seventh day of rest, for it was the very sabbath which they regarded with special reverence.[46]

In this regard, it was felt that, in times of war, inaction or indifference to enemy action should be replaced by the belief that ensuring the survival of Israel was an act of ser-

---

la clôture de la Michna." With reference to divorce, according to Gerald J. Blidstein (1991), the same diversity of opinion can be found in Hellenistic-Roman Judaism.

43 Mishnah, *'Erubin*, IV, 5. It is highly significant that, in his speech to the Jews of the Xyste of Jerusalem (Josephus, *Bell. Iud.*, II, 392), Agrippa II said that "If you observe your sabbath customs and refuse to take any action on that day, you will undoubtedly be easily defeated, as were your forefathers by Pompey, who pressed the siege most vigorously on the days when the besieged remained inactive" (τηροῦντές γε μὴν τὰ τῶν ἑβδομάδων ἔθη καὶ πρὸς μηδεμίαν πρᾶξιν κινούμενοι ῥᾳδίως ἁλώσεσθε, καθάπερ οἱ πρόγονοι Πομπηίῳ ταύτας μάλιστα τὰς ἡμέρας ἐνεργοὺς ποιησαμένῳ τῆς πολιορκίας, ἐν αἷς ἤργουν οἱ πολιορκούμενοι). See Feldman 1993, 161; Nikiprowetzky 2000, 9; Batsch 2005, 299.

44 This was, in fact, the position held by Josephus (*Ant. Iud.*, XII, 274–277; XIV, 63–64). See Batsch 2005, 279. Some Jews believed that the Law was not broken if they fought in self-defense on the day of rest, but, in practice, the decision on the defensive nature of a battle was a difficult one, as there were no hard and fast criteria according to which a struggle was equivalent to hand-to-hand fighting: ὑπὲρ τοῦ σώματος (Josephus, *Bell. Iud.*, I, 146). See Goodman and Holladay 1986, 170; Batsch 2005, 276.

45 McCasland 1943, 65; Goodman and Holladay 1986, 170; Nikiprowetzky 2000, 11 ff.

46 Josephus, *Bell. Iud.*, II, 517: Οἱ δὲ Ἰουδαῖοι κατιδόντες ἤδη πλησιάζοντα τῇ μητροπόλει τὸν πόλεμον, ἀφέμενοι τὴν ἑορτὴν ἐχώρουν ἐπὶ τὰ ὅπλα, καὶ μέγα τῷ πλήθει θαρροῦντες ἄτακτοι μετὰ κραυγῆς ἐξεπήδων ἐπὶ τὴν μάχην μηδὲ τῆς ἀργῆς ἑβδομάδος ἔννοιαν λαβόντες· ἦν γὰρ δὴ τὸ μάλιστα παρ' αὐτοῖς θρησκευόμενον σάββατον. Cf. Josephus, *Ant. Iud.*, XVIII, 318–324. See Nikiprowetzky 2000, 11. As he states in his autobiography, Josephus encountered the problem of acting against the rules on Sabbath eve faced with the threat of insurrection in Tiberias (*Vita* 159–168). The context clearly shows that this was a preventive action that was unacceptable on a Sabbath, rather than a defensive action (in *Bell. Iud.*, II, 632–637, events are described in a slightly different manner). See Nodet 1997, 72–73.

vice to God,[47] and, among others, Josephus himself[48] believed that armed struggle was entirely justified on the day of peace.[49] After all, the priests holding the Saturday service in the Temple were under threat in the name of the Sabbath itself if they chose to recklessly disregard the enemy's aggression. Thus, service to God provided an effective justification.[50] It could therefore be said that the hurdle of the Commandments (even as they remained fully in force) was overcome, in practice, on several occasions, and that certain, more liberal currents in rabbinic thought even cleared it entirely.[51]

As Étienne Nodet pointed out, in Jewish thought that emerged after the destruction of the Temple, any considerations on offensive war on Sabbath was directly or indirectly related to the notion of holy war illustrated by the campaigns of Joshua, especially in the case of the taking of Jericho, as the siege lasted seven consecutive days (*Joshua* 6).[52] In this regard, it is worth quoting a sentence attributed to Shammai the Elder, one of the foremost sages of the turn of the Era, on the minimum delay before Sabbath that should be observed before engaging in military actions:[53]

> A detachment which goes out to fight an optional war does not besiege a gentile town less than three days before the Sabbath. But if they began the siege, even on the Sabbath they do not interrupt it. And thus did Shammai the Elder expound, "[*That you may build siegeworks against the city that makes war with you,*] *until it falls* (Deut. 20:20)—even on the Sabbath."[54]

It must be acknowledged that, in practice, none of the difficulties entailed by the Sabbath Commandment prevented Jews from serving in a number of Hellenistic armies[55]

47 For instance, Josephus, *Bell. Iud.*, I, 148–150; V, 464–466. See Herr 1960–1961, 242–256 and VII; McKay 1994, 50; Nikiprowetszky 2000, 16. In any case, the weakness of the human armies had nothing to do with the divine army of the biblical tradition ... See Quine 2019.

48 For example, *Ant. Iud.*, XIII, 337; *Bell. Iud.*, I, 86; III, 532–542. See Nodet 1997, 75; Nikiprowetzky 2000, 15; Batsch 2005, 306. On the position held by the Jewish historian in the event of the war of reconquest, see Nodet 1997, 80 ff.

49 Tosefta, *'Erubin*, 3–5. Cf. *Sifre Deuteronomy* 203 and *BT*, *'Erubin* 45a. According to Robert Goldenberg (1979, 433), "While a growing number of authorities permitted military activity on the Sabbath, there continued to be those who refused to join."

50 Nikiprowetzky 2000, 16.

51 Feldman 1993, 160; Nodet 1997, 90–91; Batsch 2005, 307. On the relaxation the observance of the sabbatical year permitted by rabbis after the Bar Kochba Revolt (135 CE), see Harrington 2001, 220 note 39.

52 Nodet 1997, 87–88.

53 According to this sentence, "You must no mount an attack on a foreigner's town less than three days before the Sabbath." See Herr 1960–1961, 249–252.

54 Tosefta, *'Erubin*, 3, 7. *Cf.* *TB*, *'Erubin* 45a.

55 Batsch 2005, 250.

and, later on, in the Roman army, regardless of circumstances. As we have seen above, they served the Seleucids as mercenaries, and they were an important part of the Ptolemaic military organisation.[56] Their full integration into these armies undoubtedly prevented them from strictly observing the Sabbath rest[57] and religious scruples were almost invariably set aside during the Jewish Wars in the face of the pressing needs of the dire military situation.[58] Even considering that Jews may have exceptionally enjoyed certain privileges within Hellenistic armies, the notion that the Sabbath rest was especially respected is dubious at best. Excluding those periods in which they were exempted from military services, Jews who joined the Roman army, both willingly and by compulsion, would hardly have been able to reconcile the Sabbath rest with the demands of military life.[59] Nevertheless, there may have been exceptions for certain positions in the army which, being exempted from marches or combat, would have allowed for full observance of Jewish ritual rules. According to a legal text featured in the *Digest*, there was no shortage of such positions in the Roman army:

> Certain soldiers are granted by their conditions of service some exemption from the heavier fatigues. There are men such as surveyors, the medical sergeant, medical orderlies and dressers, ditchers, farriers, the architects, pilots, shipwrights, artillerymen, glassfitters, smiths, arrowsmiths, coppersmiths, helmet-makers, wagon-makers, roof-tile-makers, swordcutlers, water engineers, trumpet-makers, horn-makers, bow-makers, plumbers, blacksmiths, stonecutters, limeburners, woodcutters, and charcoal-burners. In the same category there are usually included butchers, huntsmen, keepers of sacrificial animals, the workshop, granary clerks, clerks responsible for monies left on deposit, clerks responsible for

56 Even if we were to assume that, during the Hellenistic period, Jews always served in separate units (Kasher 1985, 78; Trebilco 1991, 18, 187, 197–198 note 58), which is not directly supported by the sources, we cannot reconcile the requirements of military life and the Sabbath rest imposed by Judaism. On the Jewish barracks in Pelusium and Leontopolis, Josephus, *Ant. Iud.*, XIV, 99 and 131–133.

57 Nodet 1997, 67.

58 For instance, *Bell. Iud.*, II, 409–410; 450; 453; 456; 517. These passages do not contain any evidence of justification to annul the rule: Jews simply hoped that their actions in a defensive war would be met with God's approval (*Bell. Iud.*, VII, 260): see Barclay 1996, 441. On the problems caused by Jewish festivities in Jerusalem during the First Jewish War, see Price 1992, 109–114. The degree of observance of the Sabbath rest in times of war is also indicative of the various groups of insurgents (Batsch 2005, 270). Josephus (*Bell. Iud.*, II, 147) states, for instance, that the Essenes were the strictest of all Jews in their observance of the Sabbath (διαφορώτατα Ἰουδαίων ἁπάντων). In the context of the Bar Kokhba war, see Oppenheimer 2005c; Horbury 2014, 368–370; Mor 2016, 428.

59 Batsch 2005, 254.

> monies left without heirs, orderly-room staffs, grooms, horse-trainers, armoury sergeants, the herald and the trumpeter. There are all then classed as *immunes*.[60]

One member of the Dura Europos synagogue is described in an inscription as an *apothecarius*, a term which may have referred to someone "who was in charge of a military archive".[61] Moreover, we know that the standard bearers in the *Legio IV Scythica* "often served as clerks, in addition to their role in the military command," and "the large number of scribes documented in Dura has led to the assumption that every centurion was accompanied by at least one scribe-clerk who most likely was also a standard bearer".[62] It is even possible that Jews were conscripted particularly for local functions, for example, guarding the roads, and were able to live at home and do this. We know that Palmyrenes were recruited to defend the caravan routes.[63] Nevertheless, such cases remained exceptional, as the Romans generally took care not to station army units to serve in their own provinces.[64]

In any event, there is nothing in the sources available to us that would allow us to accept the theory put forward by Solomon Grayzer, according to which the high number of Jews serving the military in non-combatant posts during the 4th century was precisely the reason why the Church hierarchy made efforts to exclude them.[65] In fact, there are no reasons whatsoever to believe that Jewish soldiers were not part of combat troops. Not even the Sabbath rest was an insurmountable obstacle to their full integration in the Roman army. Jewish authors such as Philo and Josephus felt the need to justify this reality through the good will of the Roman authorities who, according to them, facilitated observance of the Sabbath rest whenever it conflicted with common military activities. Philo stated that Roman officials allowed the Jews to collect their quota of grain the next day if it fell on a Saturday[66] and Josephus mentions the

60 *Tarruntenus Paternus, Digest* 50.6.6: *Quibusdam aliquam vacationem munerum graviorum condicio tribuit, ut sunt mensores, optio valetudinarii, medici, capsarii, et artifices et qui fossam faciunt, veterinarii, architectus, gubernatores, naupegi, bllistrarii, speculari, fabri, sagittarii, aerarii, bucularum structores, carpentarii, scandularii, gladiatores, aquilices, tubarii, cornuarii, arcuarii, plumbarii, ferrarii, lapidarii, et hi qui calcem cocunt, et qui silvam infindunt, qui carbonem caedunt ac torrent. In eodem numero haberi solent lani, venatores, victimarii, et optio fabricae, et qui aegris praesto sunt, librarii quoque qui docere possint, et horreorum librarii, et librarii depositorum, et librarii caducorum, et adiutores comiculariorum, et stratores, et polliones, et custodes armorum, et praeco, et bucinator. Hi igitur omnes inter immunes habentur*. See Watson 1985, 76; Weisman 2012, 28.

61 Rosenfeld and Potchebutzky 2009, 202.

62 Rosenfeld and Potchebutzky 2009, 217.

63 Isaac 1990, 144–145.

64 Oppenheimer 2005, 191.

65 Grayzel 1968, 97. Cf. Weisman 2012, 28.

66 Philo, *Legatio ad Gaium*, 155–158.

existence of at least seven Roman laws that allowed the Jews to observe the Sabbath, exempting them from military service and from attending courts of law on that day.[67] It is true that, as Robert Goldenberg points out, "occasional attestations of the Sabbath in other pagan sources confirm that the Jewish habit was indeed well known in the Roman Empire,"[68] but the attitude of the Roman officials varied from respect and accommodation to misunderstanding and even abuse. In this regard, we cannot know which reaction prevailed in daily life, as it is difficult to distinguish between the ideals of the learned author and the practice of the Jewish community. As Fernando Bermejo Rubio aptly pointed out, "the biblical precept about the observance of Sabbath was liable to several interpretations, especially in borderline cases, arousing positions which oscillated between rigorism and laxity; after all, it involved a break in daily life, that could become annoying (Jr 17:21–27; Am 8:5). The Gospels' debates about what is allowed or not on Sabbath are confined within these intra-Jewish discussions. As happened with circumcision, some radical allegorists interpreted Sabbath in a purely symbolic manner, to such an extent that, according to Philo, they did not observe it."[69] According to the *óstraca* preserved in Edfu (*Apollinopolis Magna*), in the south of Egypt, from the period between Vespasian and Trajan (69–117 CE), local Jews paid the special tax to the *fiscus Iudaicus* even on Saturdays.[70] Considering that collecting money and writing receipts for it on the Sabbath were of course in contradiction with the rules of Jewish religion,[71] this suggests that the Jews themselves did not observe the Sabbath strictly,[72] at least when they were compelled not to do so by Roman authorities. As Willy Clarysse, Sofie Remijsen, and Mark Depauw have highlighted, "this case study of a small Jewish community in the Egyptian town Edfu shows that it was not always possible to observe the Sabbath strictly when confronted with the governmental power of the Roman Empire, despite the indications of tolerance found in Philo or Josephus. The Jews from Edfu did pay taxes on Sabbath. When they could collect the taxes themselves, however, namely in case of the Jewish tax, they took care to observe their religious practices."[73] Some texts from the Jerusalem Talmud clearly show that even there Jews were

67 Josephus, *Ant. Iud.*, XIV, 225–227 and 241–264; XVI, 167–168 and 171. See Goldenberg 1979, 415–418.

68 Goldenberg 1979, 430–436.

69 Bermejo Rubio 2020, 66.

70 Clarysse, Remijsen and Depauw 2010, 54–55; Honigman 2016, 42; Schwartz 2016, 227.

71 We know financial transactions were forbidden on Sabbath: see, for instance, Philo, *De migratione Abrahami* 91 (early 1st century CE) and Mishnah, *Shabbat*, 7, 2 (ca. 200 CE). On the frequent disregard of religious precepts (including the Sabbath) by Alexandrian Jews at the time of Josephus, see Schimanowski 2007, 133.

72 Clarysse, Remijsen and Depauw 2010, 54–56.

73 Clarysse, Remijsen and Depauw 2010, 56.

involved in economic activity on the Sabbath.[74] This shows that, as a general rule, Jews only observed the Sabbath when circumstances allowed it, and we can assume that they must have had a similar attitude when serving in the Roman army. Indeed, we have not found in the Talmudic sources any condemnation of the authorities for forcing Jewish soldiers to transgress their ancestral laws. Taking the evidence we have considered into account, army service would not pose excessive problems for Jews who were not particularly careful to observe all the Commandments (*'ammei haAretz* and others).[75]

74 Doering 1999, 387–397.

75 Oppenheimer 2005, 190–191. According to this researcher, "It should also be remembered that people who were candidates for conscription and their families had the option of freeing themselves or their sons by paying a monetary ransom, a solution which had advantages for both sides." (2005, 191).

APPENDIX 2

The Inscription of *Rufinus* the Soldier, from the Via Appia Pignatelli Catacomb (Rome)

In a research paper published in the first number of *Mitteilungen des Deutschen Archaeologischen Instituts* (*Athenische Abteilung*), Nikolaus Müller described the hypogeum of a small Roman catacomb, scarcely twenty metres away from the Via Appia Pignatelli, which he accessed for the first time on 12 April 1885. It was an *arcosolium*, which is currently inaccessible and which was almost immediately in front of the eastern atrium of the Vigna Randanini Jewish catacombs. According to his detailed description, the structure of the hypogeum was in poor condition and none of its tombs were intact, suggesting it had been violently sacked sometime after its original use. The galleries, which were crudely excavated from the granular tufa, had partially collapsed and were half covered in earth, bricks and all manner of waste. According to its discoverer, the original entrance of the catacomb was somewhere near the point from which he had accessed it, and this hypothesis was confirmed through the discovery of a flight of steps which was unearthed near there, and by the fact that both the accessible galleries and visible cubicles ended in a wall of tufa.[1]

Much of the material used to close the tombs was clearly pagan in origin (bricks stamped with figures of eagles, portrayals of the god Mercury and Victoria, as well as a number of reused inscriptions that may stretch as far back as the 2nd century CE).[2] Nevertheless, the shape of the tombs and certain peculiar features of the inscription painted on stucco that closed one of the *arcosolia*, which were very similar to the funerary and iconographic forms found in the nearby Vigna Randanini catacomb, led their discoverer to believe this was a Jewish catacomb. The oven-shaped *loculi* (*kôkhim*) in the hypogeum, which were dug perpendicularly to the wall—preferably on its lower part—that were hermetically sealed with a layer of bricks coated in stucco allowed him to draw a line between this form of burial and that observed in many tombs found in the nearby Jewish catacomb of Vigna Randanini and in all those in the Jewish catacomb of Venosa (southern Italy).[3] The inscription was painted on an *arcosolium*, in a

1 Müller 1886, 49.

2 It should not be forgotten that the funerary iconographies of the pagan Romans "form the matrix in which the funerary art of Late Antiquity—both that of the Jews and of the early Christians—germinated." (Rutgers 2013, 510).

3 Müller 1886, 51. On the features of the tombs in the Jewish catacomb of Venosa, see Colafemmina 1981.

 | DOI:10.1163/9789004507258_009

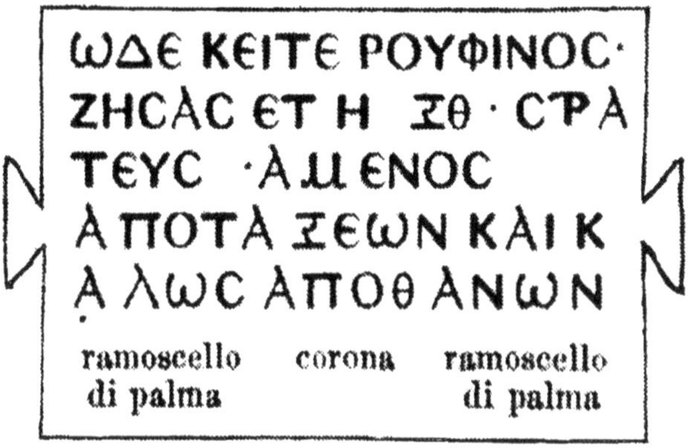

FIGURE 14 Drawing of the Rufinus inscription from the Via Appia Pignatelli Catacomb (Rome). Müller 1886, 55

tabula ansata (a tablet with dovetail handles) supported by a pair of wingless nude *putti* or cupids, at whose feet was drawn a laurel wreath surrounded by two palm branches, and it was likewise flanked on the outside by two symmetrical palm trees, one of them being complete, and the other only having part of its trunk visible. Moreover, it featured a hedgerow of reeds with a fish above as decoration between the two. The Greek letters in the highly damaged inscription are painted in white on a red background (figure 14). Above is the drawing of the inscription made by N. Müller himself.[4]

Here is the reconstruction and translation put forward by David Noy:[5]

> ὧδε κεῖτε Ῥουφῖνος | ζήσας ἔτη ξθ',
> στρα|τευσάμενος | ἀπὸ ταξέων καὶ κ|αλῶς ἀποθανών. |
> (palm) (wreath) (palm)
>
> Here lies Rufinus, who lived 69 years, having soldiered in the ranks and died well.[6]

Faced with the absence of strictly Christian signs, the catacomb's discoverer ruled out the possibility of it being a Christian cemetery from the outset. Apart from the *kôkhim*, the presence of palm branches and palm trees on the sarcophagus, which closely resembled those on the four corners of the "cubiculo delle Palme" ("Palm Tree Room") in the neighbouring catacomb of Vigna Randanini (figure 15),[7] led Nikolaus Müller to

4 Müller 1886, 55.
5 *JIWE* 2, 510–511 (No. 628). Cf. Vogelstein and Rieger 1896, I, 471 (No. 90); *CIJ* I, 52 (No. 79) and Leon 1995, 274 (No. 79).
6 *JIWE* 2, 511 (No. 628).
7 See Laurenzi 2011, 52–53.

FIGURE 15 The "Palm Tree Room" from the Vigna Randanini catacomb (Rome)
PHOTOGRAPHY BY RAÚL GONZÁLEZ SALINERO

establish a close iconographic link between this decorative motif and the palm of Palestine, which was a common motif in Jewish milieus.[8] Nevertheless, with few exceptions,[9] none of these arguments have been deemed conclusive enough by modern

8 Müller 1886, 55. Furthermore, this author establishes other stylistic relationships between the inscription and this hypogeum and the ones discovered in Venosa (with many fragments whose background had been painted red) and in Vigna Randanini (1886, 56). At least five Jewish inscriptions painted in red with the symbols of the menorah, palm tree and ethrog are attested in Rome (Mazzoleni 1975, 299).

9 Applebaum 1971, 182; Mazzoleni 1975, 298–299; Goodnick Westenholz 1994, 64. According to John Barclay (1996a, 325), "It is possible that the Rufinus whose military record was found in the Via Appia Pignatelli catacomb (*CIJ* 79) was a Jew; but there is room for doubt whether this catacomb (or this grave) was Jewish." Cf. Rodrigues 2007, 966.

historians to accept the Jewish nature of this catacomb[10] and, consequently, that of the aforementioned inscription of Rufinus.[11]

One of this theory's foremost detractors is Cinzia Vismara, according to whom the presence of hermetically closed tombs does not imply that they were necessarily Jewish as, according to her, they were not uncommon in other Christian and pagan Roman hypogeums. Although she acknowledges that the palm trees are reminiscent of the tree of Palestine that appears on Jewish coins, the author argues that this motif can also be found in other non-Jewish hypogeums such as, for instance, the catacomb of Commodilla, under the scene of the *traditio clavium*, and in the one beneath the Via Latina catacomb, in a bucolic context. Moreover, even though she admits that the formula ὧδε κεῖτε (or the more common ἐνθάδε κεῖται) is very common in Jewish Roman epigraphy, she states that examples can also be found in Christian epigraphy and in at least one pagan inscription. Furthermore, the absence of other typical symbols, such as the *menorah*, makes it difficult to identify this hypogeum as properly Jewish; to this we must add that, according to her, it is impossible for the veteran Rufinus to have been a Jewish Roman soldier due to the exemption of Jews from military service.[12] However, as we have amply demonstrated, this exemption, which was passed in specific circumstances, only had a local and temporary validity and never prevented Jews from voluntarily enlisting in the Roman army.[13] The abundant documentation available on the subject, including epigraphic sources, completely disproves this last assumption. It is true that there are grounds (mainly those stated in Vismara's detailed analysis) to challenge, at least hypothetically, the Jewish nature of this catacomb, but the exemption is not one of them.[14] Moreover, not even the remaining objections make the identification of the hypogeum of Via Appia Pignatelli as Jewish less likely than it being Christian, pagan or any other possibility. Let us see why.

Despite the fact that there are recorded cases of hermetically closed tombs in non-Jewish settings in Rome, it is undeniable that the form of burial known as *kôkh* (pl.

---

10 Ferrua 1936, 307–308; *CIJ* I, 52; Rutgers 2000, 33; van der Horst 1991, 99 note 56. On the Via Appia Pignatelli catacomb visited by Nikolaus Müller, Amy K. Hirschfeld (2008, 31) simply states that "at that time was mistakenly considered to be Jewish", without explaining why it was a mistake.

11 As Harry J. Leon (1995, 274) pointed out, "whether this is a Jewish inscription depends on whether or not the catacomb was Jewish."

12 Vismara 1986, 390–392. According to David Noy (*JIWE* 2, 181), "the assumption that the hypogeum was Jewish seems to have rested on the style of the loculus-closures and on one inscription (no.628.i) which begins ὧδε κεῖτε and has two palmbranches (identified as lulabs, but also a common motif on non-Jewish inscriptions). Four reused pagan inscriptions were also found. In fact, there seems to be no reason at all for supposing the site was Jewish, and it has been excluded by recent writers."

13 Cf. Panayotov 2004, 61.

14 González Salinero 2003, 79 note 133.

FIGURE 16 Vigna Randanini catacomb (Rome), regione B, *kôkhim* (Laurenzi 2011, 50)

kôkhim) was typically Jewish (figure 16).[15] In fact, it is abundantly attested both in Israel and the Diaspora, and it is even described in the *Mishnah*, which also prescribed its ideal dimensions.[16] It is commonly found in the Jewish catacombs of Rome,[17] and it is precisely predominant in Vigna Randanini,[18] the catacomb next to the Via Appia Pignatelli hypogeum.[19] Likewise, as Erwin Goodenough proved, we cannot deny the profoundly Jewish nature of the palm tree as a symbol, whose branches are usually associated with the *lulav* (branch or frond of a palm tree, one of the four species waved ceremonially during *Sukkot*), as can be seen in numerous examples both in Rome and in the rest of the Diaspora.[20] The two examples put forward by Cinzia Vismara as proof

15 In the words of Danilo Mazzoleni (1987, 310), oven-shaped tombs or *kôkhim* are "di tradizione eminentemente giudaica."

16 Vitale 1986, 177–178.

17 Mazzoleni 1975, 290; Vitale 1986, 182–183.

18 Mazzoleni 1975, 296; Green 2008, 152.

19 The admittedly sporadic presence of amphora burials in the hypogeum of Via Appia Pignatelli is not incompatible with the *khôkim*, as this burial form was used by Jews and non-Jews alike. In fact, the amphora burials were discovered in the Monteverde Jewish catacomb (Rutgers 2000, 60).

20 Goodenough 1953–1965, II, 40, 43–44, 54, 58–62, 64–65, 68–69, 104, 106, 110–112, 115, 125–126, 215, 217–218, 255, 257, 259–260, 262, 266–267, 287, 289; IV, 75, 144, 145–147, 156, 162, 165–166. On the palm tree as a Jewish symbol, see Goodenogh 1953–1965, II, 20–21, 31, 34,

that the palm tree was also present in non-Jewish contexts are, in my opinion, unconvincing owing to their exceptional nature. The first one has to do with a Christian representation of the *traditio clavium* in a clearly antiquarian connection with a scene set in a Palestinian context,[21] whereas the second one is related to a pastoral scene depicting varied and exotic flora that has nothing to do with the iconographic context of the painted inscription of Rufinus, in which the palm tree is the unique symbolic motif.

With regards to the additional decoration accompanying the inscription, the hedgerow of reeds with a fish above it is certainly unique, but the fish itself is extremely important in Vigna Randanini, and appears to be one of the most important symbols in Judaism.[22] Moreover, according to Erwin Goodenough, "the crown of laurel, while a commonplace in the period, is quite of a piece with Vigna Randanini, and with the Jewish catacombs of Torlonia in Rome and at Gammarth in Tunisia."[23]

As regards the text of the inscription, even though the formula ὧδε κεῖτε does appear in Christian epigraphy,[24] the absence of other clues related to the Christian funerary world makes it unlikely that this hypogeum was Christian. This conventional formula is a very familiar form on Jewish tombstones.[25] Furthermore, we must keep in mind that the name *Rufinus* (Ῥουφῖνος) is well attested in Roman Jewish epigraphy, and we therefore cannot rule out the possibility that he was, indeed, a Jew.[26]

Moreover, the location and ground plan have not been taken sufficiently into account in the analysis of the relationship between the hypogeum found in Via Appia

---

56, 62, 64, 66, 94, 114, 147; IV, 73, 132, 140, 145, 149, 150, 163–164. Cf. Frankel and Platkin Teutsch 1995, 125; Varon 2001, 275; Laurenzi 2011, 53.

21 Apart from its close relationship with the victory of the martyrs, the palm tree was also used by Christians in Biblical settings. See Baudry 2009, 98–100.

22 Goodenough 1953–1965, II, 34.

23 Goodenough 1953–1965, II, 34–35.

24 Vismara 1986, 502, note 308. An isolated occurrence in a pagan setting (*IGVR* II, 756) can safely be attributed to external influences.

25 In Rome: *JIWE* 2, No. 71 (= *CIJ* I, No. 357); No. 237 (= *CIJ* I, No. 129); No. 238 (= *CIJ* I, No. 147); No. 239 (= *CIJ* I, No. 167); No. 337 (= *CIJ* I, No. 120); No. 459 (= *CIJ* I, No. 25); No. 463 (= *CIJ* I, No. 55); No. 488 (= *CIJ* I, No. 37); No. 578 (= *CIJ* I, No. 510); No. 584 (= *CIJ* I, No. 537). See Fasola, 1976, 45. In Venosa: *JWIE* I, No. 46 (= *CIJ* I, No. 588); No. 47 (= *CIJ* I, No. 579); No. 48 (= *CIJ* I, No. 594); No. 50 (= *CIJ* I, No. 577); No. 52 (= *CIJ* I, No. 576); No. 66 (= *CIJ* I, No. 591); No. 76 (= *CIJ* I, No. 600); No. 77 (= *CIJ* I, No. 578); No. 104; No. 114 (= *CIJ* I, No. 619b); No. 115 (= *CIJ* I, No. 619c); No. 116 (= *CIJ* I, No. 619d). See also Goodenogh 1953–1965, II, 34; Varon 2001, 275.

26 *JIWE* 2, No. 198 (= *CIJ* I, No. 480), from Monteverde catacomb; No. 414 and No. 501 (both from Villa Torlonia catacomb); *Cf. JIWE* 2, No. 12, from Lorium (La Bottaccia, Lazio). See Ilan 2008, III, 538–539. Its feminine form (*Rufina*), *JIWE* 2, No. 145 (= *CIJ* I, No. 31), from Monteverde catacomb; and No. 609 (= *CIJ* I, No. 9*), from Vigna Randanini catacomb. See Ilan 2008, III, 611–612.

Pignatelli and the nearby Jewish catacomb of Vigna Randanini.[27] In this regard, it is worth keeping in mind that the hypogeum discovered by Nikolaus Müller was scarcely twenty metres away from the main entrance of the Vigna Randanini catacomb and that, like it, it also featured remnants of pagan materials from a previous structure that was used for funerary purposes from the Severan period onwards (early 3rd century CE), when these vast tracts of land were acquired by the Jewish community.[28] With regards to the large stretch of terrain (ca. 18,000 m^2) in which the Vigna Randanini catacomb was located, Silvia Cappelletti stated that "[...] the area was occupied by non funerary structures built at the end of the second century, as some clear traces in the building of via Appia Pignatelli indicate. As from the Severian period, certainly in the third century, the area was run by the Jewish community that restored and modified the buildings on the surface and begun the excavation of the galleries along a north-east and south-west axis that led naturally towards via Appia [...]. The north-east gallery of this second catacomb includes a hypogeum, probably of pagan origin, and opens a new area with a large presence of cubicula and with narrow galleries, an area that has not been explored because it had already collapsed at the time of the first excavations. The entrance stairway to this area has not been found, but, as we have already remarked, the area has totally collapsed. Unlike the fortuitous crossing with the gallery dug from via Appia Pignatelli, this gallery makes a 90 degree corner and stretches to the area with the painted rooms. Most probably, the diggers of the Jewish catacomb, who had taken over the management of the pagan hypogeum later, tried to reach and incorporate this new area that they decided to use as a burial place without religious bias. The site known today as the catacomb of Vigna Randanini originally was probably made of three distinct areas. Considering that the paintings of cubicola 13 and 14 date back to 220/250 CE, the incorporation of the hypogeum was possibly decided after the end of the third century [...]."[29] Thus, there are reasonable grounds to draw a close link between the structure of the Vigna Randanini catacomb currently known to us and the hypogeum discovered by Nikolaus Müller at Via Appia Pignatelli, and this would likewise allow us to confirm the 3rd century dating put forward by him for this funerary area and the inscription of Rufinus featured on one of its *kôkhim*.[30]

Thus, the features of the underground cemetery at Via Appia Pignatelli are much closer to the Jewish cultural sphere than to the Christian or pagan ones, and we can safely assume that Rufinus the veteran was a Jewish soldier from the 3rd century CE.

27 Lucrezia Spera (1999, 367) concludes that the land between via Appia and via Appia Pignatelli was the site of a "highly developed" necropolis characterized by monumental sepulchers of the first through fourth centuries CE.

28 Mazzoleni 1976; *Idem* 1987, 310; Cappelletti 2006, 157.

29 Cappelletti 2006, 158–159. Cf. Laurenzi 2011, 51–52.

30 Müller 1886, 55. Danilo Mazzoleni (1975, 299) already pointed out that the dating suggested by Müller was "accettabile" (cf. Goodenogh 1953–1965, II, 32.)

APPENDIX 3

A Critical Rereading of the Inscription of Flavia Optata Found in Concordia

When I published the paper on Jewish military service in the Roman army 2003, I unthinkingly accepted the reading and interpretation of this inscription put forward by most scholars. The inscription (figure 17), which was found in the military cemetery (necropoli di Levante) of *Concordia Sagittaria*, features Flavia Optata as the wife, daughter, or even mother of a Jewish soldier who had allegedly served in a *numerus* known as the *Regii Emeseni Iudaei* (the "Royal Emesene Jews").[1] Most scholars accepted Theodor Mommsen's authoritative opinion,[2] which has been uncritically reproduced by the latest scientific literature.[3] I must admit that, at the time, I did not pay enough attention to Michael P. Speidel's revealing research paper (1996) in which he made a convincing argument against the Jewish nature of this inscription since he initially started from the uncertain reading made by its discoverers[4] and that I accepted the opinion of its latest editor, David Noy, according to whom "this interpretation seems much more unsatisfactory than the 'Jewish' one". Nevertheless, Noy himself also cast doubt on the subject when he stated that "Flavia Optata's connection with a Jewish army unit is far from certain."[5] In fact, regarding this inscription, Roger Tomlin went as far as to state that "Mommsen's restoration and expansion of the reading are both far-fetched."[6] A few years later, Walter Ameling delved deeper into this uncertainty, pointing out that "it is neither proved nor even probable that Flavia Optata and her husband were Jews."[7] Nevertheless, he did not expand on the specific reasons that made him doubt its Jewish nature beyond the fact that this inscription had no "connection with any kind of epigraphic habit in Syria." We may assume that, had he been aware of the aforementioned study by Michael P. Speidel, he would have had solid arguments

1 González Salinero 2003, 63–71.

2 *CIL* V, 1064 (No. 8764).

3 Hoffmann 1963, 50–51; Cracco Ruggini 1964, 932, n. 21; Applebaum 1970, 4; Hoffmann 1970, I, 60; Applebaum 1971, 182; Tomlin 1972, 270; *Idem* 1976, 190; Pavan 1987, 20–21; Williams 1998, 97; Castritius 2003b, 5–6; Oppenheimer 2005, 187; Schoenfeld 2006, 123 (= 2019, 103–104); Kovács 2010, 175; Rocca 2010, 26–27; Lightstone 2011, 196; Weisman 2012, 26–27; Olshanetsky 2018, 18–19; Guichaoua 2019, 25, 53, and 213; Roux 2020, 30.

4 Berlini 1874, 290–291; Bertolini and Henzer 1874, 29 (No. 23) and 42, n. 15. Cf. Lefort 1875, 344.

5 *JIWE* 1, 11.

6 Tomlin 1998, 49, n. 116.

7 Ameling 2009, 223, n. 50.

 | DOI:10.1163/9789004507258_010

FIGURE 17 Inscription of Flavia Optata (*Concordia*). Portogruaro. Museo Nazionale Concordiese (inv. nº 289)
PHOTOGRAPHY BY RAÚL GONZÁLEZ SALINERO (DECEMBER 2019)

available to bolster his scepticism. This is what Marco Rocco did a few years later in his brilliant analysis of the *auxilia palatina* in the structure of the Late Roman army.[8]

The inscription is reproduced below:

Transcription:

FLAVIA OPTATA MILI DE
NUM REGI EMESIV DE
R V SI QUIS POS OVIT
ME ARC VOLU AP ENFE
R VI AUR LIB VNA[9]

Here is David Noy's restoration, following Th. Mommsen's interpretation:

Flavia Optata mili(tis) de | num(ero) regi(oru) Emes(enoru) Iude(o)|ru. si quis pos ovitu | me(u) arc(a) volu(erit) ap(erire), en(feret) fi(sci) | ⟨vir⟩(ibus) aur(i) lib(ra) una. (lulav?).

8 Rocco 2012, 163–164.

9 The first transcription suggested by its discoverors (Bertolini 1874, 290; Bertolini and Henzen 1874, 29) and which was reproduced by Lefort (1875, 344) was slightly different: *FLAVIA OPTATA MILI · DE | NVM · REGI · EMESIVDE | RV SI QVIS POSOVITV | ME · ARCVOLVAPEN · FI | RVI · AVR · LIB · VNA.*

> Flavia Optata, (wife?) of a soldier from the troop of the royal Emesene Jews. If anyone after my decease wants to open the tomb, s/he will pay to the resources of the treasury one pound of gold.[10]

The first lines of the epitaph feature a typical funerary formula, followed by a legal clause establishing a fine for whoever should desecrate the tomb (these are conventional formulas, of which there are many more examples within the necropolis in which this document was found).[11] The use of the first person by Flavia Optata in the epitaph can be explained by the absence of the soldier, probably her husband, whose name may have been omitted by accident, as in other inscriptions from Concordia.[12] The funerary fine makes sense in the absence of next of kin, as the grave would then be looked after by the city treasury, which would likewise collect any penalties arising from its desecration.[13] The epigraph has been dated to the second half of the fourth century or the early 5th century, when the name *Flavius* was particularly common among soldiers.[14] We must keep in mind that both their wives and their children[15] were deemed to be members of the same military units in which they served, as is the case here.[16]

If we were to accept the traditional reading of the inscription, as it was established by its latest editor in *JIWE 1* (1993) and *IJO III* (2004), we would also have to acknowledge that it is the only proof available of the existence of a military unit named *Regii Emeseni Iudaei*. We do know that Emesa, *Augusta Libanensis* from the 4th century onwards (modern-day Homs), was a city in Syria with a significant Jewish community.[17] According to Jerome,[18] its bishop, Eusebius, wrote against that community in

10 *JIWE 1*, No. 6 (= *CIL* V, No. 8764 = *ILCV*, No. 4857 = *CIJ I*, No. 640 = Lettich 1983, No. 57 = *EDCS* 5401563 = *IJO III*, 68). Instead of identifying it as a *lulav* (as David Noy suggests in his restoration), the simplified figure in the shape of an ear of corn at the end of the epigraph seems to be a simple palm tree, whose presence—as with the *hedera*—can be found in inscriptions of this nature—especially from Christian contexts—in Concordia as well (Lettich 1983, 98; Mazzoleni 2002, 174, fig. 88; Cuscito 2009, 164).

11 Impallomeni 1984, 121–132; Cresci Marrone 2001, 246; Mazzoleni 2002, 176–178. On funerary fines in general, see Rossi 1975; Carletti 2008, 100–103.

12 See Lettich 1983, n[os] 63, 65, 69, 72, 76, 80. A reference to the mother of a soldier from Salona is attested in Concordia (Lettich 1983, No. 71). *Cf. CIL* III, 2208 (*EDCS* 27700029).

13 Impallomeni 1984, 125–126.

14 *JIWE 1*, 9.

15 Several epitaphs in the Concordia military cemetery feature the names of soldiers alongside those of their wives or brides (Lettich 1983, n[os] 1, 13, 14, 19, 34, 40, 62, 64, 67, 70, 74, 75, 79).

16 See Hoffman 1963, 50; Cresci Marrone 2001, 246. Cf. Jones 1992, 630–631.

17 See David Noy and Hanswulf Bloedhorn in *IJO III*, 69–70.

18 Jerome, *De uir. ill.*, 91.

the mid-4th century. We could therefore reasonably assume that the Jewish soldiers that made up the aforementioned unit would have been drawn from the inhabitants of that city, if only initially. It is true that it is not recorded under this complete name in the *Notitia Dignitatum* (only *Regii*), but this hurdle was temporarily cleared by arguing that its western section was only drafted after the year 421,[19] by which time the unit would have already been disbanded or reorganized under a different name and conditions, as the ban on Jewish military service in the Roman army would have come into force by then.[20] According to the supporters of this traditional reading of the inscription, a connection could be drawn between the abbreviated mention of the alleged military unit and the *Regii* featured in the *Notitia Dignitatum*.[21] In its eastern section they are recorded among the *auxilia palatina*,[22] and in the western one, as one of the *legiones comitatenses* stationed in Italy under the command of the *Magister Peditum Praesentalis*.[23] A number of military units (*numeri*) of *Emeseni*, some of which may hypothetically have been exclusively made up from Jews,[24] are known to us of epigraphic sources, but there is no conclusive evidence on the subject.

According to Ralf Scharf, the name *Regii Emeseni Iudaei* could plausibly have originated from the large number of Jews present in this alleged military unit from the time of its creation. We must keep in mind that, as with other detachments stationed in Concordia, the last epithet in unit nomenclature is, in fact, its main name, which in this case would be *Iudaei*.[25] Constantin Zuckerman has argued that this was clearly a homogenous group of soldiers in the auxiliary forces of the Roman army.[26] According to him, there is sufficient evidence (which he fails to provide) to believe that it was a unit raised by emperor Aurelian circa 273, after Palmyrene power had been broken, from the Jews in the region of Emesa, then part of the province of *Syria Phoenicia*, and that their peculiar name paid homage to the former priest-kings of the city. Nevertheless, by this logic,

19 Neira Faleiro 2005, 42–43.

20 Scharf 1997, 358–359.

21 Neither do Shi'mon Applebaum's (1971, 182) arguments in favour of a different reading of the abbreviation *regi(orum)* as *regi(onis)* hold up in this regard, as they were based on the assumption that the royal house of Emesa had been dethroned by Domitian and that the first *numeri* did not appear until the reign of Hadrian. Thus, according to this researcher, the real issue would be to pinpoint the location of the *Regio Iudaeorum Emesenorum*. However, he fails to provide any explanations on the origins of the alleged unit of Emesene Jews, nor does he consider the possibility that, due to its peculiar composition, it retained its original name when it became a *numerus* in the Roman army.

22 *Not. Dig. Orientis*, VI, 49 (Neira Faleiro 2005, 191).

23 *Not. Dig. Occidentis*, V, 229 and VII, 32 (Neira Faleiro 2005, 341 and 350). See Tomlin 1976, 190.

24 See *JIWE 1*, 10. Cf. Woods 1992, 404.

25 Scharft 1997, 354.

26 Zuckerman 1993, 19.

this unit may have also been created at an earlier date. In fact, the alleged eastern origin of the *Regii Emeseni Iudaei* and the fact that the inhabitants of Emesa were not entitled to Roman citizenship, at least until the Severan period, would explain the fact that they were part of the *auxilia palatina*.[27] Shim'on Applebaum pointed out that *Cohors I milliaria Emesenorum sagittariorum equitata c. R.* was stationed at the camp of Intercisa from the late 2nd century CE onwards, and that there was a synagogue outside the camp whose *archisynagogus* was a *praepositus stationis* at the time of Severus Alexander.[28] Since it was assumed that there was a Jewish unit whose names included that of *Emeseni*, Applebaum felt he could posit a relationship (the nature of which he could not specify) between both units, considering that the presence of a synagogue in its *uicus* meant that the cohort of Intercisa may have had Jews among its ranks. However, there is no evidence to support Shim'on Applebaum's assumption that this region (*regio*) may have been the original recruiting ground of the hypothetical *numerus Emesenorum Iudaeorum* and that it moved from there to Concordia in the 5th century alongside other *numeri* from Dalmatia and Germania.[29]

We likewise cannot accept the hypothesis put forward by David Woods according to which the Jewish soldiers who assaulted the church of Theonas, in Alexandria, in 356, were the *Regii*:[30] for one thing, there is not a single source that mentions the name of this auxiliary unit in relations to the events and, for another, we have no means of drawing a link between the Jewish troops that stormed the church, which were unambiguously described as legionaries by Athanasius[31] (λεγεώνων στρατιωτῶν) and which were then serving in Egypt under the *dux Syrianus*, and the *Regii*, who were operating in Gaul at the time. There is no evidence that they were temporarily stationed in the eastern front, as David Woods baselessly supposed.[32]

This entire structure of historical arguments based on the inscription of Flavia Optata that acts as the basis for the alleged existence of a *numerus* that was, at least initially, entirely made up from Jewish soldiers, must inevitably come crashing down when faced with the new reading of the inscription by Michael P. Speidel. The key to its interpretation lies in the correct restoration of the abbreviations in the second and third lines of which hinges on the meaning of the series of letters that make them up: EMESIV DE R V SI.

---

27 Scharf, 1997, 357.
28 Applebaum 1971, 183.
29 Applebaum 1971, 183.
30 Woods 1992.
31 Athanasius, *Hist. Ar. ad mon.*, 81, 7; *Apol. pro fuga sua*, 24; *Apol. ad Const. imp.*, 25.
32 Woods 1992, 407. The complete and detailed development of my argument in González Salinero 2003, 70–71. Cf. also Scharft 1997, 347.

As we pointed out earlier, the latest editor essentially accepted the restoration put forward by Theodor Mommsen in 1877: *Emes(enorum) Iude(o)ru(m)*.[33] However, he put the letter *O*, which had been artificially added by Mommsen when faced with the difficulty of recognizing the Latin word *Iudaeus* in its genitive plural form (*Iudaeorum*), between parentheses. Michael P. Speidel, starting from the reading of *EME* as an abbreviation of the verb *emo* in the third person singular of the present perfect and of *SIV* as the initial letters of the reflexive pronoun (*sibi*), as early editors did,[34] restored the text as follows: *emi(t) sib(i) de r(e) v(iri)*. That is, Flavia Optata bought the coffin for herself from her husband's money. As the author aptly points out, we must keep in mind that "in twenty-seven of thirty-seven cases, military owners of coffins from this graveyard state that they bought the coffin with their own money (*arcam emit sibi de proprio*), hence Optata, too, is likely to have done so."[35] The oddities of her spelling and wording parallel other Concordia coffins: *sivi* for *sibi*,[36] *eme* for *emit*,[37] and the phrase *de rem sua*.[38] All these oddities are regular vulgarisms of late Latin.[39] Formulas such as *de proprio*,[40] *de proprio suo*,[41] *de proprio labore suo*,[42] or *de proprio meo*[43] are all very common in Concordian epigraphy. *Vir* is traditionally abbreviated to *v* in phrases such as *v(ir) e(gregius)* or *v(ir) c(larissimus)*.[44] Thus, and even though it does not follow the proper form of text restoration according to the rules of epigraphy, the alternative put forward by Speidel (1996, 164), with its orthographic corrections, is clearly much more plausible:

Flavia Optata, mili(tis) de
num(ero) Regi(orum), emi(t) sib(i) de
r(e) v(iri). Si quis pos(t) obit(um)
me(um) arc(am) volu(erit) ap(erire) infe-
r(at) (fisci) vi(ribus) aur(i) lib(ram) una(m).

---

33 *JIWE* 1, 6 (= *IJO* III, 68).
34 Bertolini 1874, 290; Bertolini and Henzer 1874, 42; Lefort 1875, 344.
35 Speidel 1996, 164.
36 Lettich 1983, nº 48.
37 Lettich 1983, nºs 8 (*emerat*), 25 (*emerunt*) 35 (*emit*), 44 (*emet*), 45 (*emi*), 46 (*emit*), 70 (*emerunt*), 74 (*em*), 78 (*emit*), 79 (*emerunt*).
38 Lettich 1983, nº 14.
39 Precisely, epigraphic documentation from Concordia, where the majority of the 21 military units (*numeri*) attested there were made up from barbarian contingents (Batavi, Heruli, Bructeri …), displays a progressive vulgarisation of the Latin language. See Cresci Marrone 2001, 246 and 249.
40 Lettich 1983, nºs. 24, 48, 52, 72, 74, 104.
41 Lettich 1983, nºs. 9, 12, 18, 22, 23, 35, 36, 38, 43, 46, 47, 51, 68, 69, 76, 79.
42 Lettich 1983, nºs. 21, 34, 41, 52, 53.
43 Lettich 1983, nºs 20, 45.
44 Lettich 183, nº 17. See Speidel 1996, 164 note 6.

I have restored the text in accordance with the proper rules of epigraphy below, without altering its reading and, therefore, its translation:

> *Flavia Optata, mili(tis) de*
> *num(ero) Regi(orum), em{e}⸢i⸣ (t) si{v}⸢b⸣ (i) de*
> *r(e) v(iri). Si quis pos⟨t⟩ o{v}⸢b⸣ịṭ(um)*
> *me(um) arc(am) volu(erit) ape(rire) {e}⸢i⸣nfe*
> *r(at) (fisci) vi(ribus) aur(i) lib(ram) una(m)* ❦ ⋘

> Flavia Optata, (wife) of a soldier
> of the Regii unit, bought (this) for herself from
> her husband's means. If anyone after my death
> opens the coffin, he shall pay
> the fisc one pound of gold.

This rereading of this inscription would rule out the possibility that the *numerus* known as the *Regii* was made up from Jewish soldiers and refute the epithets that had been unduly tacked onto its name (*Emeseni Iudaei*). However, the origins of this auxiliary unit, which is simply mentioned as *Regii* in a number of Late Antique sources, remain a mystery. Based on a passage of the *Epitome de Caesaribus* (41, 3), Michael Speidel argued that these *Regii*—"the king's men"—were originally the Alamanni commanded by king Crocus who made a significant contribution to Constantine's imperial proclamation in 306 after his father, Constantius Clorus, passed away in York.[45] As auxiliary household troops, Crocus' men surely were the *auxilium palatinum* of the *Regii*.[46] A few decades later, another unit bearing this name appears alongside the Batavi, which were part of the field army in Gaul, and they even served under the orders of Caesar Julian in the battle of Strasbourg (*Argentorate*) in August 357, in which the Alamanni were defeated and repulsed beyond the Rhine after their invasion of part of the Empire.[47] According to Constantin Zuckerman, the presence of the *Regii* (which were still, according to him and the vast majority of authors who have covered the subject, identified under the extended name of *Regii Emeseni Iudaei*) in these events is crucial evidence that they could not have been a legion, but rather, an auxiliary unit. In obedience to the

45 Speidel 1996, 165. Cf. Rocco 2012, 495.

46 *Not. Dig. Orientis*, VI, 49 (Neira Faleiro 2005, 191). See Speidel 1996, 165; Rocco 2012, 281, n. 96.

47 Ammianus Marcellinus, XVI, 12, 45. See Hoffmann 1969–1970, I, 68; Pavan 1987, 21; Woods 1992, 404–405; Speidel 1996, 166; Nicasie 1998, 228; Rodríguez González 2001, II, 531; Rocco 2012, 163 note 157. We do not know if this auxiliary unit still predominantely recruited Alamanni at the time. If it were so, the *Regii* would have had to fight their former compatriots on that occasion.

usual order of battle of Roman armies at the time, according to which the legions brought up the rear while the auxiliaries led the van, it is inconceivable that the *Regii*, being deployed in the vanguard, were legionaries. This is likewise confirmed by the fact that they acted alongside other units, such as those made up from the Batavi (who are also present in inscriptions from Concordia under the name *numerus Batavorum seniorum*),[48] who have been unmistakably identified as auxiliaries.[49]

A number of plausible conclusions may be drawn from this careful analysis. Firstly, there is no evidence that conclusively proves that the *Regii* ever bore the name *Emeseni Iudaei* and that they were therefore Jewish troops, even when the unit was first raised. Secondly, as per the sources available to us (Ammianus Marcellinus, *Notitia Dignitatum Orientis et Occidentis* and now the inscription of Flavia Optata), the only epithet known to have been attached to this unit, which belonged to the *auxilia palatina* and would later become a legionary force, was that of *Regii*. Regardless of the fact that its soldiers were most likely first drafted from among the Alamanni—as can be deduced from the anonymous *Epitome de Caesaribus*—, according to the sources, the *Regii* shared a close bond with the Batavi, a *numerus* that is amply attested in the epigraphy from the same military cemetery of Concordia in which the inscription of Flavia Optata was discovered.

48 Lettich 1983, n°s 26, 35–41. See Speidel 1996, 166–167. According to Roger Tomlin (1972, 270 note 6), "the *Batavi seniores* and the *Regii* occupy complementary positions in the lists of *Or*. 5 and *Or*. 6."

49 Zuckerman 1993, 19. On this subject, see also Rodríguez González 2001, II, 531.

Prosopographic Map

1. *Zamaris* (Jewish commander). Batanaea (Bashan, Déraa), Trachonitis, Syria. 10–7 BCE.
2. *Rufus* (Jewish commander?). Land of Israel. Late first century BCE.
3. *Gratus* (Jewish commander?). Land of Israel. Late first century BCE.
4. *Volumnius* (Jewish commander?). Land of Israel. Late first century BCE.
5. *Aebutius* (Jewish decurion?). Simonias (west of Nazareth), Judaea. 66–67 CE.
6. 4,000 Jewish soldiers to fight brigands. Sardinia Island. 19 CE.
7. Troopers of Zamaris. Batanaea (Bashan, Déraa), Syria. Until 132 CE.
8. *Jacimus*, son of Zamaris (Jewish commander). Bathyra (district of Trachonitis), Syria. Beginning of the First century CE.
9. *Philip*, son of Jacimus (Jewish commander). Bathyra (district of Trachonitis), Syria. First century CE.
10. *Archieus* (Jewish centurion). Hauran (Auranitis, Jebel al-ʿArab), Syria. 108 CE.
11. *Chares* (Jewish commander). Eithae, Batanaea (Bashan, Déraa), Trachonitis, Syria. Beginning of the first century CE.
12. *Tiberius Julius Alexander* (Jewish commander). Armenia. 63 CE.
13. *Antiochus* (Jewish renegade soldier). Antioch (Turkey), Syria. 67 CE.
14. *Matthaius*, son of Polaius (veteran Jewish soldier). Misenum (Italy). 68 CE.
15. *Agrippa Apamma* (veteran Jewish soldier?). Misenum (Italy). 68 CE.
16. *Gallus* (Jewish commander?). Judaea. First century CE.
17. *Marcus Dama Surus Garasenus* (veteran Jewish soldier). Ravenna (Italy). 71 CE.
18. *Lucius Cornelius Simon* (veteran Jewish soldier?). Ravenna (Italy). 71 CE.
19. *Titus Mucius Clemens* (Jewish commander). Dora/Dor (Bir el-Malik, Israel), Syria-Phoenice. Between 50 and 90 CE.
20. *Iesous*, son of Papios (Jewish soldier?). Apollinopolis Magna (Edfu), Egypt. Between 71 and 121 CE.
21. *Gaius Iulius Apollinarius* (Jewish soldier?). Karanis (near Fayyum), Egypt. 107 CE.
22. *Aninios* (Jewish centurion). Apollinopolis Magna (Edfu), Egypt. 116 CE.
23. Jewish soldiers. Athribic district, Egypt. 130 CE.
24. *Barcallipus*, son of Aggaeus (veteran Jewish soldier). Samosata (Samsat, Turkey), Syria. (144 CE).
25. *Barsimsus Callisthenis* (Jewish soldier). Tibiscum (Jupa, Caraș-Severin, Romania), Upper Dacia. 157 CE.
26. *Abdior* (Jewish soldier). Egypt (provenience unknown). Late second century CE.
27. *Lucius Maecius Archon* (Jewish centurion?), Rome (Italy). Second-Third centuries CE.
28. *Barichius* (Jewish soldier?). Oxyrhynchus (El-Bahnasa, district of Minia), Egypt. 205 CE.

 | DOI:10.1163/9789004507258_011

29. *Malochus* (Jewish soldier?). Oxyrhynchus (El-Bahnasa, district of Minia), Egypt. 205 CE.
30. *Cosmius* (Jewish officer). Intercisa (Dunaújváros-Dunapentele, Hungary), Pannonia. 222–234 CE.
31. *M. Aurelius Malchias*, son of Mocur (Jewish groom-soldier of the consular office). Intercisa (Dunaújváros-Dunapentele, Hungary), Pannonia. Third century CE.
32. *Mocur*, father of *M. Aurelius Malchias* (Jewish *signifer*). Intercisa (Dunaújváros-Dunapentele, Hungary), Pannonia. Third century CE.
33. *M. Aurelius Sallumas* (veteran Jewish soldier). Intercisa (Dunaújváros-Dunapentele, Hungary), Pannonia. Third century CE.
34. *Germanus Valens* (Jewish soldier). Intercisa (Dunaújváros-Dunapentele, Hungary), Pannonia. Third century CE.
35. *Aurelius Eliaseir* (Jewish officer?). Intercisa (Dunaújváros-Dunapentele, Hungary), Pannonia. Third century CE.
36. *Aurelius Damas* (Jewish soldier). Intercisa (Dunaújváros-Dunapentele, Hungary), Pannonia. Third century CE.
37. *Germanos*, son of Isak the Palmyrene (Jewish soldier). Beth She'arim (Besara, Tivon, Israel), Galilea. Third century CE.
38. *Rufinus* (Jewish soldier). Rome (Italy). Third century CE.
39. *Joses* (Jewish officer and archisynagogus). Oescus (Gigen, Bulgary), Upper Moesia. Third-Fourth centuries CE.
40. *Flavius Euphorbius* (Jewish soldier?). Oescus (Gigen, Bulgary), Upper Moesia. Fourth century CE.
41. *Flavius Valentianus* (Jewish soldier?). Oescus (Gigen, Bulgary), Upper Moesia. Fourth century CE.
42. *Sanbatio* (Jewish soldier?). Oescus (Gigen, Bulgary), Upper Moesia. Fourth century CE.
43. *Agrius Evangelus* (Jewish soldier). Rome (Italy). Third-Fourth centuries CE.
44. *Reginus* (Jewish soldier). Rome (Italy). Third-Fourth centuries CE.
45. Antiquarian soldier (Jewish soldier). Rome (Italy). Beginning of the Fourth century CE.
46. *Flavius Eusebius* (Jewish officer). Aphrodisias (Turkey), Asia Minor. Beginning of the Fourth century CE.
47. *Heron* (Jewish soldier). Dyonisias (northwest corner of the Fayyum), Egypt. Ca. 350 CE.
48. Jewish troopers under *dux Syrianus*. Alexandria, Egypt. 356 CE.
49. *Benjamin the Centenarius* (Jewish centenarius). Ioppe (Jaffa), Palestine. Fourth century CE.
50. *Abraam* (Jewish soldier). Magdolum (Migdol, Tell el-Herr, Nile Delta), Egypt. Fourth century CE.

51. *Moses* (Jewish soldier). Magdolum (Migdol, Tell el-Herr, Nile Delta), Egypt. Fourth century CE.
52. *Eusebius* (Jewish *agens in rebus*). Ioppe (Jaffa), Palestine. Fourth-Fifth centuries CE.
53. *Aurelius Isaac* (Jewish soldier?). Hermopolis Magna (El Ashmunein, Minya), Egypt. 507 CE.
54. *Flavius Silvanus* (Jewish soldier?). Hermopolis Magna (El Ashmunein, Minya), Egypt. 507 CE.
55. *Flavius Serapion* (Jewish soldier?). Hermopolis (El Ashmunein, Minya), Egypt. 507 CE.
56. *Yudan*, son of Shimon (Jewish soldier). Meroth (Northern Galilee). Sixth century CE.
57. *Flavius Jacob* (Jewish soldier?). Syene and Philae (Aswan), Egypt. 573 CE.
58. *Flavius Abraham* (Jewish soldier?). Syene and Philae (Aswan), Egypt. 573 CE.
59. *Stercorius* (Jewish soldier?). Syene and Philae (Aswan), Egypt. 573 CE.
60. *Flavius Abraham* (Jewish soldier?). Syene and Philae (Aswan), Egypt. 573 CE.

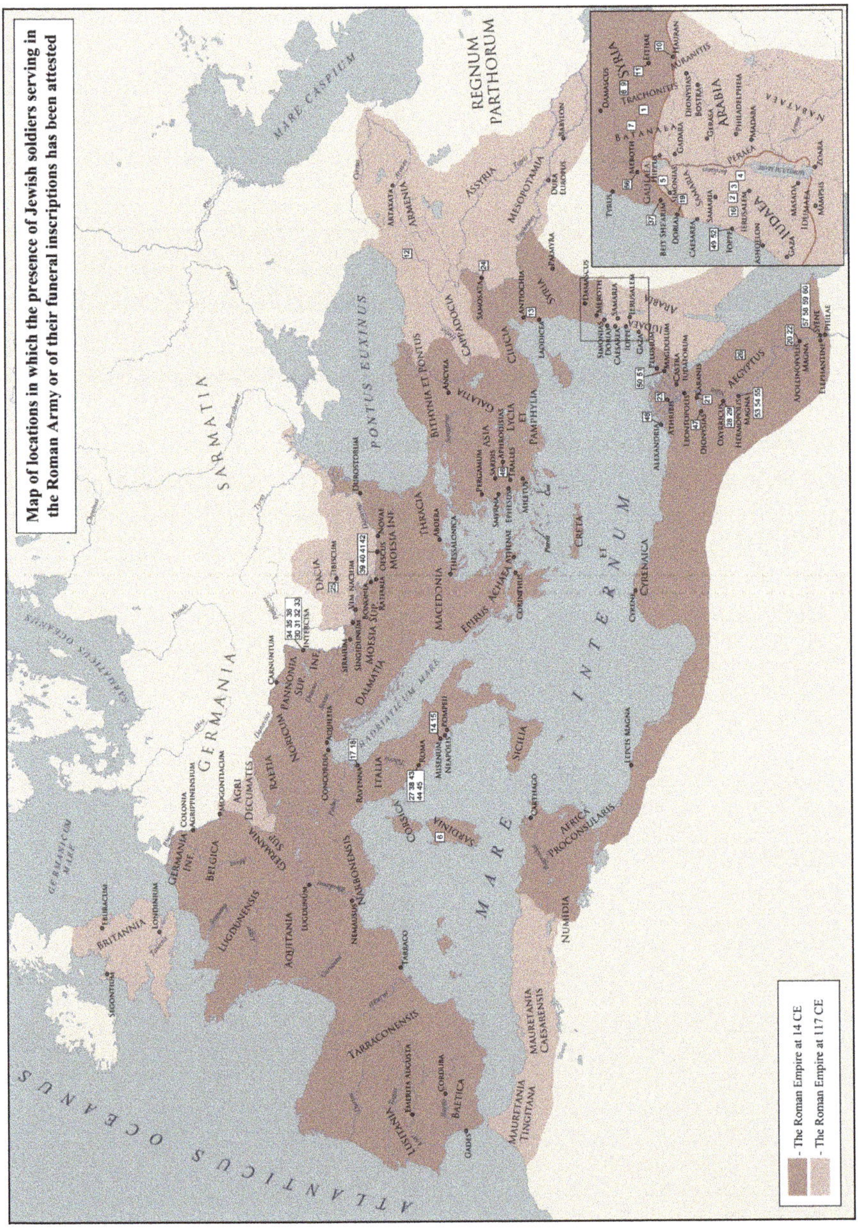
Map of locations in which the presence of Jewish soldiers serving in the Roman Army or of their funeral inscriptions has been attested
- The Roman Empire at 14 CE
- The Roman Empire at 117 CE

Sources

Graeco-Roman Literature

Ammianus Marcellinus, *Rerum gestarum libri qui supersunt*: ed. W. Seyfarth, L. Jacob-Karau, and I. Ulmann (1978), *Rerum gestarum libri qui supersunt*, Leipzig: B.G. Teubner; ed. and transl. John C. Rolfe (1935–1940), *Ammianus Marcellinus in Three Volumes*, Cambridge, Mass./London: William Heinemann/Harvard University Press (Loeb Classical Library 300, 315 and 331).

Appian, *Historia romana*: ed. and transl. by H. White (1912–1913), *Appian's Roman History in Four Volumes*, London / Cambridge (Mass.): William Heinemann / Harvard University Press (Loeb Classical Library 2, 3, 4, 5).

Arrian, *Alexandri anabasis*: ed. and transl. by E.I. Robson (1966–1967), *Arrian. Anabasis Alexandri*, London / Cambridge (Mass.): William Heinemann / Harvard University Press (Loeb Classical Library 236, 269).

Arrian, *De acie contra alanos*: ed. and transl. J.G. DeVoto (1993), *Flavius Arrianus, Tactical Handbook and The Expedition against the Alans*, Chicago: Ares Publishers.

Caesar, *Commentarii belli Gallici*: ed. and transl. by H.J. Edwards (1919), *Caesar. The Gallic War*, London/New York/Cambridge (Mass.): William Heinemann/G.P. Putnam's Sons/Harvard University Press (Loeb Classical Library 72).

Cassius Dio, *Historiae Romanae*: ed. and transl. H.B. Foster and E. Cary (1914–1927), *Dio's Roman History*, London/Cambridge (Mass.), William Heinemann/Harvard University Press (Loeb Classical Library 32, 37, 53, 66, 82–83, 175–177).

Cicero, *Epistulae ad familiares*: ed. by D.R. Shackleton Bailey (1988), *M. Tullius Cicero. Epistulae ad Familiares*, Leipzig: B.G. Teubner; transl. by P.G. Walsh (2008), *Cicero. Selected Letters*, Oxford: Oxford Universuty Press.

Diodorus of Sicily, *Bibliotheca historica*: ed. and transl. by C.H. Oldfather *et alii* (1933–1984), *Diodorus of Sicily in Twelve Volumes*, London / Cambridge (Mass.): William Heinemann / Harvard University Press (Loeb Classical Library 279, 303, 340, 375, 384, 389, 399, 409, 422–423).

Frontinus, *Strategemata*: ed. and transl. by Ch.E. Bennet (1929), *The Stratagems and the Aqueducts of Rome*, London/New York: William Heinemann/G.P. Putnam's Sons (Loeb Classical Library 174), 3–327.

Livy, *Ab urbe condita*: ed. and transl. by B.O. Foster (1967–1987), *Livy in Fourteen Volumes*, Cambridge (Mass.)/London: Harvard University Press/William Heinemann (Loeb Classical Library 114, 133, 172, 191, 295, 301, 313, 332, 354–355, 367, 381, 396, 404).

Martial, *Epigrammata*: ed. and transl. by W.C.A. Ker and D.R.Sh. Bailey (1919–1920 and 1993), *Martial, Epigrams*, London/New York: William Heinemann/G.P. Putnam's Sons/Harvard University Press (Loeb Classical Library 94a, 95a and 480).

Pliny the Younger, *Epistulae*: ed. and transl. by W. Melmoth (1915), *Pliny, Letters*, London/New York: William Heinemann/The Macmillan Co. (Loeb Classical Library 55 and 56).

Plutarch, *De superstitione*: ed. and transl. by F.C. Babbitt (1928), *Plutarch's Moralia in Fifteen Volumes*, Cambridge (Mass.)/London: Harvard University Press/William Heinemann (Loeb Classical Library 222), 451–495.

Rutilius Namatianus, *De reditu suo*: ed. J. Vessereau and F. Préchac (1961), *Rutilius Claudius Namatianus, Sur son retour*, Paris: Les Belles Lettres; transl. M. Malamud (2016), *Rutilius Namatianus' Going Home. De reditu suo*, Abingdon (Oxon)/New York: Routledge.

Scriptores Historiae Augustae: ed. and. transl. D. Magie (1922–1932), *The Scriptores Historiae Augustuae in Three Volumes*, London/New York/Cambridge (Mass.): William Heinemann/G.P. Putnam's Sons/Harvard University Press (Loeb Classical Library 139–140, 263).

Seneca, *De constantia sapientis*: ed. and transl. by J.W. Basore (1928), *Seneca. Moral Essays*, London/New York: William Heinemann/G.P. Putnaman's Sons (Loeb Classical Library 214).

Strabo, *Geographica*: ed. and transl. by H.L. Jones (1917–1967), *The Geography of Strabo in Eight Volumes*, London / Cambridge (Mass.): William Heinemann / Harvard University Press (Loeb Classical Library 49–50, 182, 196, 211, 223, 241, 267).

Suetonius, *De vita Caesarum libri VIII*: ed. M. Ihm (1908), *Suetonius. De vita Caesarum libri VIII*, Leipzig: B.G. Teubner; transl. by C. Edwards (2000), *Suetonius. Lives of the Caesars*, Oxford: Oxford University Press.

Vegetius, *Epitoma rei militaris*: ed. M.D. Reeve (2004), *Vegetius. Epitoma rei militaris*, Oxford: Oxford University Press (Scriptorum Classicorum Bibliotheca Oxoniensis); transl. N.P. Milner (1993), *Vegetius. Epitome of Military Science*, Liverpool: Liverpool University Press (Translated Texts for Historians 16).

Servius Honoratus (Grammaticus), *Commentarius in Vergilii opera* (*in Aeneida*): ed. G. Thilo and H. Hagen (1881–1902), *Servii Grammatici qui feruntur in Vergilii Carmina commentarii*, Leipzig: B.G. Teubner (5 vols.).

Stern, Menahem (1976–1984), *Greek and Latin Authors on Jews and Judaism*, Jerusalem: The Israel Academy of Sciences and Humanities (3 vols.).

Velleius Paterculus, *Historiae Romanae*: ed. and transl. F.W. Shipley (1961), *Velleius Paterculus. Compendium of Roman History*, William Heinemann / Harvard University Press (Loeb Classical Library 152).

Tacitus, *Annales*: ed. H. Heubner (1994), *Cornelii Taciti libri qui supersunt*, Stuttgart: B.G. Teubner (*editio correctior*); transl. by J.C. Yardley (2008), *Tacitus. The Annals. The Reigns of Tiberius, Claudius, and Nero*, Oxford: Oxford University Press.

Tacitus, *Historiae*: ed. and transl. by J. Jackson (1962), *Tacitus. The Histories*, London/Cambridge (Mass.): William Heinemann/Harvard University Press (Loeb Classical Library) (orig. 1925).

Zosimus, *Nova Historia*: ed. L. Mendelssohn (1887), *Zosimi comitis et exadvocati fisci Historia nova*, Leipzig: B.G. Teubner; transl. J.J. Buchanan and H.T. Davis (1967), *Zosimus. Historia nova: The Decline of Rome*, San Antonio (Texas): Trinity University Press; transl. R.T. Ridley (1982), *Zosimus. New History*, Canberra: Australian Association for Byzantine Studies.

Christian Literature

Athanasius, *Historia Arianorum ad monachos*: ed. H.-G. Opitz (1940), *Athanasios Werke, II, 1, 1: Die Apologien*, Berlin, Walter de Gruyter, 183–230; transl. Ph. Schaff and H. Wace (1892), *St. Athanasius: Select Works and Letters*, New York: The Christian Literature Company (A Select Library of Nicene and Post-Nicene Fathers, Second Series, vol. IV), 270–302; transl. R. Flower (2016), *Imperial Invectives against Constantius II. Athanasius of Alexandria, History of Arians. Hilarius of Poitiers, Against Constantius. Lucifer of Cagliari, The Necessity of Dying for the Son of God*, Liverpool: Liverpool University Press (Translated Texts for Historians 67).

Athanasius, *Apologia pro fuga sua*: ed. J.M. Szymusiak (1987), *Athanase d'Alexandrie. Deux apologies. À l'empereur Constance. Pour sa fuite*, Paris: Les Éditions du Cerf (SC 56 bis), 133–167; transl. Ph. Schaff and H. Wace (1892), *St. Athanasius: Select Works and Letters*, New York: The Christian Literature Company (A Select Library of Nicene and Post-Nicene Fathers, Second Series, vol. IV), 255–265.

Athanasius, *Apologia ad Constantium imperatorem*: ed. J.M. Szymusiak (1958), *Athanase d'Alexandrie. Deux apologies. A l'empereur Constance. Pour sa fuite*, Paris: Les Éditions du Cerf (SC 56), 88–132; transl. Ph. Schaff and H. Wace (1892), *St. Athanasius: Select Works and Letters*, New York: The Christian Literature Company (A Select Library of Nicene and Post-Nicene Fathers, Second Series, vol. IV), 238–253.

Augustine of Hippo, *Sermones Dolbeau*: ed. François Dolbeau (1996), *Augustin d'Hippone. Vingt-Six sermons au peuple d'Afrique*, Paris: Études Augustiniennes.

Augustine of Hippo, *Sermones Mayence*: ed. François Dolbeau (2009), *Augustin d'Hippone. Vingt-Six sermons au peuple d'Afrique retrouvès à Mayence*, Paris: Études Augustiniennes.

Historia acephala: ed. A. Martin (1985), *Histoire "Acephale" et Index Syriaque des lettres festales d'Athanase d'Alexandrie*, Paris: Les Éditions du Cerf (SC 317).

Jerome, *Commentarius in Isaiam*: ed. R. Gryson *et alii*, *Commentaires de Jérôme sur le Prophète Isaïe*, Freiburg: Herder, 1993–1999 (5 vols.); ed. M. Adriaen (1963), *S. Hieronymi presbyteri opera*, Turnhout: Brepols (*Corpus Christianorum. Series Latina* 73 and 73a); transl. Th.P. Scheck (2015), *St. Jerome: Commentary on Isaiah*, New York/Mahwah: The Newman Press.

Jerome, *De viris illustribus*: ed. D. Hurst and M. Adriaen (1969), *S. Hieronymi presbyteri opera*, Turnhout: Brepols (*Corpus Christianorum. Series Latina* 77), 392–449; transl.

Th.P. Halton (1999), *Saint Jerome: On Illustrious Men*, Washington, D.C.: Catholic University of America Press.

John Chrysostom, *Commentarium in Isaiah*: ed. J. Dumortier (1983), *Jean Chrysostome. Commentaire sur Isaïe*, Paris: Les Éditions du Cerf (SC 304).

John Malalas, *Chronica*: ed. J. Thurn and M. Meier (2009), *Johannes Malalas, Weltchronik*, Stuttgart: Hiersemann, 2009; transl. E. Jeffreys *et alii* (1986), *The Chronicle of John Malalas*, Melbourne: Australian Association for Byzantine Studies.

Lucifer of Cagliari, *De Athanasio* (*Quae absentem nemo debet iudicare nec damnare*): ed. G.F. Diercks (1978), *Quae absentem nemo debet iudicare nec damnare sive De Athanasio*, Turnhout: Brepols (*Corpus Christianorum. Series Latina* 8), 3–132.

Eusebius of Caesarea: *Chronicon* [Latin translation by Jerome]: ed. R. Helm (1956), *Eusebius Werke: siebenter Band. Die chronik des Hieronymus* (*Hieronymi Chronicon*), Berlin: Akademie.

Orosius, *Historiarum adversus paganos libri VII*: ed. M.-P. Arnaud-Lindet (1990–1991), *Orose*, Paris: Les Belles Lettres (3 vols.); transl. A. Fear (2010), *Orosius. Seven Books of History against the Pagans*, Liverpool: Liverpool University Press (Translated Texts for Historians 54).

Procopius of Caesarea, *De aedificiis*: ed. and transl. H.B. Dewing (1940), *Procopius. On Buildings*, London/Cambridge, MA: William Heinemann/ Harvard University Press (Loeb Classical Library 343).

Pseudo-Joshua the Stylite, *Chronica*: ed. and transl. F.R. Trombley and J.W. Watt (2000), *The Chronicle of Pseudo-Joshua the Stylite*, Liverpool: Liverpool University Press (Translated Texts for Historians 32).

Sulpicius Severus, *Chronica*: ed. P. Parroni (2017), *Sulpicii Severi Chronica*, Turnhout: Brepols (*Corpus Christianorum. Series Latina* 63); transl. Ph. Schaff and H. Wace (1894), *Sulpicius Severus, Vincent of Lerins, John Cassian*, New York: The Christian Literature Company (A Select Library of Nicene and Post-Nicene Fathers, Second Series, vol. XI), 71–122; transl. Richard J. Goodrich (2015), *Sulpicius Severus. The Complete Works*, New York: The Newman Press (orig. 1962).

Tertullian, *De corona*: ed. by E. Kroymann (1954), *Tertulliani Opera*, in *Corpus Christianorum. Series Latina*, 2, Turnhout: Brepols, 1039–1065; transl. by R.D. Sider (2001), *Christian and Pagan in the Roman Empire. The Witness of Tertullian*, Washington, D.C.: Catholic University of America Press, 115–136.

Theodoret of Cyrus, *Commentarius in Isaiam*: ed. J.-N. Guinot (1980), *Théodoret de Cyr. Commentaire sur Isaïe*, Paris: Les Éditions du Cerf (SC 276), I, 208–210.

Zacharias of Mytilene, *Historia ecclesiastica*: ed. and Latin transl. E.W. Brooks (1924), *Historia Ecclesiastica Zachariae Rhetori vulgo adscripta*, Paris/Louvain: E Typographeo Reipublicae/E typographeo I.-B. Istas (*Corpus Scriptorum Christianorum Orientalium. Scriptores Syri* 87 and 88).

Epigraphy, Papyrology and Ostraca

AE = *L'Année Épigraphique* (1888–), Paris: Ernest Leroux and PUF.

Beševliev, Veselin Ivanov (1952), *Epigrafski prinosi* (*Epigraphic Contributions*), Sofia: Bulgarian Academy of Sciences.

BGU IV = (1912), *Aegyptische Urkunden aus den Königlichen Museen zu Berlin. Griechische Urkunden*, IV, Berlin: Weidmann.

Campbell, Brian (1994), *The Roman Army, 31 BC–AD 337: A Sourcebook*, London/New York: Routledge.

CIG = Böckh, Augustus *et alii* (1977), *Corpus Inscriptionum Graecarum*, Hildesheim: Georg Olms (4 vols.) (orig. Berlin: Officina Academica, 1828–1877).

CIIP I,1 = Cotton, Hannah M. *et alii* (2010), *Corpus Inscriptionum Iudaeae / Palaestinae. A Multi-Lingual Corpus of the Inscriptions from Alexander to Muhammad, I,1. Jerusalem (1–704)*, Tübingen: Walter de Gruyter.

CIIP I,2 = Cotton, Hannah M. *et alii* (2012), *Corpus Inscriptionum Iudaeae / Palaestinae. A Multi-Lingual Corpus of the Inscriptions from Alexander to Muhammad, I,2. Jerusalem (705–1120)*, Tübingen: Walter de Gruyter.

CIIP II = Ameling, Walter *et alli* (2011), *Corpus Inscriptionum Iudaeae / Palaestinae. A Multi-Lingual Corpus of the Inscriptions from Alexander to Muhammad, II. Caesarea and the Middle Coast (1121–2160)*, Tübingen: Walter de Gruyter.

CIIP III = Ameling, Walter *et alii* (2014), *Corpus Inscriptionum Iudaeae / Palaestinae. A Multi-Lingual Corpus of the Inscriptions from Alexander to Muhammad, III. South Coast (2161–2648)*, Tübingen: Walter de Gruyter.

CIJ I = Frey, Jean-Baptiste (1936), *Corpus Inscriptionum Iudaicarum. Recueil des inscriptions juives qui vont du IIIe siècle avant Jésus-Christ au VIIe siècle de notre ère, I. Europe*, Città del Vaticano/Roma: Pontificio Istituto di Archeologia Cristiana.

CIJ II = Frey, Jean-Baptiste (1952), *Corpus Inscriptionum Iudaicarum. Recueil des inscriptions juives qui vont du IIIe siècle avant Jésus-Christ au VIIe siècle de notre ère, II. Asie-Afrique*, Città del Vaticano/Roma: Pontificio Istituto di Archeologia Cristiana.

CIJG = Battinou, Zanet (2018), *Corpus Inscriptionum Judaicarum Graeciae. Corpus of Jewish and Hebrew Inscriptions from Mainland and Island Greece (Late 4th Century B.C.E.–15th Century)*, Athens: The Jewish Museum of Greece.

CIL = *Corpus Inscriptionum Latinarum* (1863–), Berlin/New York: Georgium Reimerum/Walter de Gruyter.

Cotton, Hannah M. (1995), "The Archive of Salome Komaïse Daughter of Levi: Another Archive from the 'Cave of Letters'," *Zeitschrift für Papyrologie und Epigraphik* 105: 171–208.

Cowley, Arthur Ernest (1923), *Aramaic Papyri of the Fifth Century B.C.*, Oxford: Clarendon Press.

CPJ I = Tcherikover, Victor A. and Alexander Fuks (1957), *Corpus Papyrorum Judaicarum (volume I)*, Cambridge (Mass.): Harvard University Press.

CPJ II = Tcherikover, Victor A. and Alexander Fuks (1960), *Corpus Papyrorum Judaicarum (volume II)*, Cambridge (Mass.): Harvard University Press.

CPJ III = Tcherikover, Victor A., Alexander Fuks, Menahem Stern, and David M. Lewis (1964), *Corpus Papyrorum Judaicarum (volume III)*, Cambridge (Mass.): Harvard University Press.

DDbDP = *The Duke Databank of Documentary Papyri* (https://papyri.info).

DMIPERP = Zeichmann, Christopher B., *Database of Military Inscriptions and Papyri of Early Roman Palestine* (http://www.armyofromanpalestine.com).

Dobó, Arpadus (1940), *Inscriptiones extra fines Pannoniae Daciaeque repertae ad res earundem provinciarum pertinentes*, Budapest: A Pázmány Péter Tudományegyetem.

EDCS = *Epigraphische-Datenbank Clauss-Slaby* (http://www.manfredclauss.de).

Erdélyi, Gizella and Ferenc Fülep (1954), "Katalog der Steindenkmäler," in L. Barkóczi *et alii*, *Intercisa I (Dunapentele-Sztálinváros). Geschichte der Stadt in der Römerzeit*, Budapest: Akadémiai Kiadó, 277–332.

Ewing, Walter (1895), "Greek and Other Inscriptions Collected in the Hauran," *Palestine Exploration Fund. Quarterly Statement* 27 (January): 41–60, 131–160, 265–280 and 346–354.

Grelot, Pierre (1972), *Documents araméens d'Égypte*, Paris: Éditions du Cerf.

Hunt, Arthur Surridge and Campbell Cowan Edgar (1932–1934), *Select Papyri. Private and Public Documents*, Cambridge (MA)/London: Harvard University Press (Loeb Classical Library 266 and 282).

Kraeling, Emil G. (1953), *The Brooklyn Museum Aramaic Papyri*, New Haven: Yale University Press.

Kraemer, Ross Shepard (2004), *Women's Religions in the Greco-Roman World. A Sourcebook*, Oxford: Oxford University Press.

IGLS XVI = Sartre, Maurice and Annie Sartre-Fauriat (2020), *L'Auranitide (1. Qunawat et la bordure nord du Jebel al-'Arab; 2. Suwaydā' et la bordure ouest du Jebel al-'Arab)*, Beyruth: Institut Français du Proche-Orient (2 vols.).

IGR = Cagnat, René (1964) *Inscriptiones Graecae ad res Romanas pertinentes*, Roma: "L'Erma" di Bretschneider (orig. Paris: Leroux, 1906–1927).

IGVR = Moretti, Luigi (1968–1990), *Inscriptiones Graecae Vrbis Romae*, Roma: Istituto italiano per la Storia Antica (5 vols.).

IJO I = Noy, David, Alexander Panayotov and Hanswulf Bloedhorn (2004), *Inscriptiones Judaicae Orientis, I. Eastern Europe*, Tübingen: Mohr Siebeck.

IJO II = Ameling, Walter (2004), *Inscriptiones Judaicae Orientis, II. Kleinasien*, Tübingen: Mohr Siebeck.

IJO III = Noy, David and Hanswulf Bloedhorn (2004), *Inscriptiones Judaicae Orientis, III. Syria and Cyprus*, Tübingen: Mohr Siebeck.

ILBulg = Gerov, Boris, *Inscriptiones Latinae in Bulgaria repertae*, Sofia: Kliment Ohridski, 1989. [https://epigraphy.packhum.org/book/192?location=12].

ILJug = Sasel, Anna and Jaro (1963), *Inscriptiones latinae quae in Iugoslavia inter annos MCMXL et MCMLX repertae et editae sunt*, Ljubljana: Musei Nationalis Labacensis.

ILCV = Diehl, Ernest (1925–1931), *Inscriptiones Latinae Christianae Veteres*, Berlin: Weidmann (3 vols.) [*Supplementum*: ed. J. Moreau and H.I. Marrou (1985), Zürich: Weidmann].

ILS = Dessau, Hermann (1892–1916), *Inscriptiones Latinae Selectae*, Berlin: Weidmann.

JIGRE = Horbury, William and David Noy (1992), *Jewish Inscriptions of Graeco-Roman Egypt*, Cambridge: Cambridge University Press.

JIWE 1 = Noy, David (1993), *Jewish Inscriptions of Western Europe, Volume 1. Italy (Excluding the City of Rome), Spain and Gaul*, Cambridge: Cambridge University Press.

JIWE 2 = Noy, David (1995), *Jewish Inscriptions of Western Europe, Volume 2. The City of Rome*, Cambridge: Cambridge University Press.

Lettich, Giovanni (1983), *Le iscrizioni sepolcrali tardoantiche di Iulia Concordia*, Trieste: Centro Studi Storico-Religiosi Friuli-Venezia Giulia.

OGIS I = Dittenberger, Wilhelm (1903), *Orientis Graeci Inscriptiones Selectae*, volumen prius, Leipzig: S. Hirzel.

Ostracon Clermont-Ganneau 152: ed. Dupont-Sommer, A. (1949), "L'ostracon araméen du Sabbat," *Semitica* 2: 29–39.

P.Abinn. = ed. Bell, H.I., V. Martin, E.G. Turner and D. van Berchem (1962), *The Abinnaeus Archive. Papers of a Roman Officer in the Reign of Constantius II*, Oxford: Clarendon Press (nos. 1–82).

P.Lond. II = ed. Kenyon, Frederic G. (1898), *Greek Papyri in the British Museum. Catalogue, with Texts, II*, London: British Museum (nos. 139–484).

P.Mich. VII = ed. Sanders, Henry Arthur and James Eugène Dunlap (1947), *Michigan Papyri, VII. Latin Papyri in the University of Michigan Collection*, Ann Arbor: University of Michigan Press.

P.Mich. VIII = ed. Youtie, Herbert Chayyim and John Garrett Winter (1951), *Michigan Papyri, VIII. Papyri and Ostraca from Karanis*, Ann Arbor: University of Michigan Press.

P.Mich. IX = ed. Husselman, Elinor M. (1971), *Michigan Papyri, IX 29. Papyri from Karanis (Third Series)*, Cleveland: The Press of Case Western Reserve University.

P.Oxy. = ed. Grenfell, Bernard P. and Arthur S. Hunt (1898–1999), *The Oxyrhynchus Papyri*, London: The Egypt Exploration Fund, 1898–2006 (70 vols.)

P.Tebt. = ed. Grenfell, Bernard P., Arthur S. Hunt and J. Gilbart Smyly (1902–1938), *The Tebtunis Papyri*, London: Henry Frowde.

P.Yadin (*P.Babatha*) = ed. and transl. Lewis, Naphtali, Yigael Yadin and Jonas C. Greenfield (1989), *The Documents from the Bar Kokhba Period in the Cave of Letters: Greek Papyri*, Jerusalem: Israel Exploration Society / Hebrew University; ed. and transl. Yadin, Yigael, Hannah Cotton and Andrew Gross (2002), *The Documents from the Bar Kokhba Period in the Cave of Letters: Hebrew, Aramaic, and Nabatean-Aramaic*

Papyri, Jerusalem: Israel Exploration Society; transl. Kraemer, Ross Shepard (2004), *Women's Religions in the Greco-Roman World. A Sourcebook*, Oxford: Oxford University Press, 143–151.

Porten, Bezalel (2011), *The Elephantine Papyri in English: Three Millennia of Cross-cultural Continuity and Change* (ed. by Joel Farber *et alii*), Atlanta: Society of Biblical Literature (orig. Leiden: E.J. Brill, 1996).

Rahmani, Levi Yitzaq (1994), *A Catalogue of Jewish Ossuaries in the Collections of the State of Israel*, Jerusalem: The Israel Antiquities Authority/The Israel Academy of Sciences and Humanities.

RGDS = *Res Gestae Divi Saporis* (inscription at Naqsh-e Rustam): ed. and French transl. Maricq, André (1958), "*Classica et Orientalia*, 5. *Res Gestae Divi Saporis*," *Syria* 35 (3–4): 295–360.

RIU = Barkóczi, László and András Mócsy (1971–), *Die römischen Inschriften Ungarns*, Amsterdam: Hakkert.

RMD = Roxan, Margaret and Paul A. Holder (1954–2006), *Roman Military Diplomas*, London: University College London Institute of Archaeology (5 vols.).

Roueché, Charlotte (1989), *Aphrodisias in Late Antiquity: The Late Roman and Byzantine Inscriptions*, London: Society for the Promotion of Roman Studies (expanded and revised version in http://insaph.kcl.ac.uk/ala2004/index.html).

Schwabe, Moshe and Baruch Lifshitz (1974), *Beth Sche'arim. The Greek Inscriptions*, New Brunswick/Jerusalem Rutgers University Press/Israel Exploration Society/Hebrew University.

Scheiber, Alexander (1983), *Jewish Inscriptions in Hungary. From the 3rd Century to 1686*, Budapest/Leiden: Akadémiai Kiadó/E.J. Brill.

SEG = *Supplementum Epigraphicum Graecum*, Alphen aan den Rijn/ Amsterdam/Leiden: A.W. Sijthoff/J.C.Gieben/E.J. Brill, 1923– (= Brill on line).

ŠKZ = Inscription from *The Kaabah of Zoroaster*: ed. Hackl, Ursula, Bruno Jacobs, and Dieter Weber (2010), *Quellen zur Geschichte des Partherreiches: Textsammlung mit Übersetzungen und Kommentaren*, Göttingen: Vandenhoeck & Ruprecht, II, 573–587 (transl. J. Nabel in *Parthian Sources Online*: http://parthiansources.com).

TAD 1 = Porten, Bezalel and Yardeni (1999), *Textbook of Aramaic Documents from Ancient Egypt, 1. Letters*, Winona Lake: Eisenbrauns.

Vandorpe, Katelijn, Willy Clarysse and Herbert Verreth (2015), *Graeco-Roman Archives from the Fayum*, Leuven: Peeters.

Yon, Jean-Baptiste and Julien Aliquot (2016), *Inscriptions grecques et latines du Musée national de Beyrouth*, Beyrouth: Ministère de la Culture.

Jewish and Graeco-Jewish Sources

BT = *The Babylonian Talmud*: transl. Isidore Epstein (ed.) (1961), *The Babylonian Talmud*, London: Soncino Press (18 vols.).

Bible: *The New English Bible* (1970), Oxford/Cambridge: Oxford University Press/Cambridge University Press (orig. 1961).

Buber, Salomon (1885), *Midrasch Tanchuma. Ein agadischer Commentar zum Pentateuch von Rabbi Tanchuma ben Rabbi Abba I–III*, Wilna: Wittwe & Gebrüder Romm (English transl. *Sefaria. A Living Library of Jewish*: https://www.sefaria.org).

Flavius Josephus, *Antiquitates Iudaicae*: ed. and transl. by H.St.J. Thackeray (1956–1961), *Josephus. Jewish Antiquities*, London/Cambridge (Mass.): William Heinemann/Harvard University Press (Loeb Classical Library 203, 210); transl. by L.H. Feldman (2000), *Flavius Josephus. Judean Antiquities 1–4*, Leiden/Boston: Brill; transl. St. Mason *et alii* (2000–), *Flavius Josephus. Translation and Commentary. Jewish Antiquities*, Leiden/Boston: Brill (7 vols.).

Flavius Josephus, *Contra Apionem*: ed. and transl. by H.St.J. Thackeray (1926), *Josephus with an English Translation in Eight Volumes, I. Life and Against Apion*, London/New York: William Heinemann/G.P. Putnam's Sons (Loeb Classical Library 186), 161–411; transl. by J.M.G. Barclay (2007), *Flavius Josephus. Translation and Commentary, vol. 10. Against Apion*, Leiden/Boston: E.J. Brill.

Flavius Josephus, *De bello Iudaico*: ed. and transl. by H.St.J. Thackeray (1961–1965), *Josephus. The Jewish War*, London/Cambridge (Mass.): William Heinemann/Harvard University Press (Loeb Classical Library 242, 281, 326, 365, 410, 433, 456); transl. St. Mason *et alii* (2008), *Flavius Josephus. Translation and Commentary, vol. 1B. Judean War 2*, Leiden/Boston: Brill.

Flavius Josephus *Vita*: ed. and transl. by H.St.J. Thackeray (1926), *Josephus with an English Translation in Eight Volumes, I. Life and Against Apion*, London/New York: William Heinemann/G.P. Putnam's Sons (Loeb Classical Library 186), 1–159; transl. St. Mason (2001), *Flavius Josephus. Translation and Commentary, vol. 9. Life of Josephus*, Leiden/Boston/Köln: Brill.

JT = *The Jerusalem Talmud*: transl. Jacob Neusner (ed.) (2009), *The Jerusalem Talmud. A Translation and Commentary*, Peabody (Massachusetts): Hendrickson Publishers Marketing (CD-ROM) (orig. Chicago: University of Chicago Press, 1982–1994, 35 vols.)

Letter of Aristeas: *Aristeae ad Philocratem epistula cum ceteris de origine versionis LXX interpretum testimoniis*: ed. A. Pelletier (1962), *Lettre d'Aristée a Philocrate*, Paris: Les Éditions du Cerf (Sources Chrétiennes 89); transl. B.G. Wright III (2015), *The Letter of Aristeas. 'Aristeas to Philocrates' or 'On the Translation of the Law of the Jews'*, Berlin/Boston: Walter de Gruyter.

1Maccabees: "The First Book of the Maccabees," in *The Revised English Bible with the*

Apocrypha (1989), Oxford/Cambridge: Oxford University Press/ Cambridge University Press, 155–183.

II Maccabees: "The Second Book of the Maccabees," in *The Revised English Bible with the Apocrypha* (1989), Oxford/Cambridge: Oxford University Press/ Cambridge University Press, 184–205.

Midrash Aggadat Bereshit: transl. L.M. Teugels (2001), *Aggadat Bereshit*, Leiden/Boston/Köln: Brill.

Midrash Genesis Rabbah: transl. H. Freedman (1939), *Midrash Rabbah, I. Genesis*, London: The Soncino Press (2 vols.).

Mishnah: transl. Jacob Neusner (1988), *The Mishnah: A New Translation*, New Haven (Conn.): Yale University Press.

Philo of Alexandria, *De somniis*: ed. and transl. F.H. Colson and G.H. Whitaker (1988), *Philo in Ten Volumes (and Two Supplementary Volumes)*, Cambridge (Mass.) / London: Harvard University Press / William Heinemann (Loeb Classical Library 275), V, 283–579.

Philo of Alexandria, *Legatio ad Gaium*: ed. and transl. by E.M. Smallwood (1970), *Philonis Alexandrini Legatio ad Gaium*, Leiden: Brill (orig. 1961).

Tosefta: transl. by J. Neusner *et alii* (1977–1986), *The Tosefta Translated from the Hebrew*, Hoboken (New Jersey): Ktav Publishing House Inc. (6 vols.)

Graeco-Roman Legal and Official Texts

CIust. = *Codex Iustinianus*: ed. P. Krüger (1954), *Corpus Iuris Civilis, vol. III. Codex Iustinianus*, Berlin: Weidmann.

CTh. = *Codex Theodosianus*: ed. Th. Mommsen (^{4}1971), *Theodosiani libri XVI cum constitutionibus Sirmondianis*, Berlin, Weidmann (orig. 1902); transl. Cl. Pharr (1952), *The Theodosian Code and Novels and the Sirmondian Constitutions*, Princeton: Princeton University Press.

Delmaire, Roland, Jean Rougé, and François Richard (2005), *Les lois religieuses des Empereurs romains de Constantin à Theodose II (312–438), vol. I. Code Théodosien, livre XVI*, Paris, Les Éditions du Cerf.

Delmaire, Roland, Jean Rougé, Olivier Huck, François Richard, and Laurent Richard (2009), *Les lois religieuses des Empereurs romains de Constantin à Theodose II (312–438), vol. II. Code Théodosien, livres I–XV, Code Justinien, Constitutiones Sirmondiennes*, Paris, Les Éditions du Cerf.

Digest = *Digesta Iustiniani*: ed. Th. Mommsen and P. Krüger (1954), *Corpus Iuris Civilis, vol. II. Digesta Iustiniani*, Berlin: Weidmann; transl. A. Watson (1998), *The Digest of Justinian*, Philadelphia: University of Pennsylvania Press (4 vols.).

Itinerarium Antonini: ed. O. Cuntz (1990), *Itineraria Romana, vol. 1. Itineraria Antonini Augusti et Burdigalense*, Leipzig: B.G. Teubner (orig. 1929).

Linder, Amnon (1987), *The Jews in Roman Imperial Legislation*, Detroit (Michigan)/Jerusalem: Wayne State University Press/The Israel Academy of Sciences and Humanities.

Magnou-Nortier, Élisabeth and Michel Rouche (2002), *Le Code Théodosien. Livre* XVI *et sa réception au Moyen Âge*, Paris: Les Éditions du Cerf.

Notitia Dignitatum: ed. C. Neira Falero (2005), *La* Notitia Dignitatum. *Nueva edición crítica y comentario histórico*, Madrid: CSIC (Col. Nueva Roma, 25); and O. Seeck ([3]1983), *Notitia Dignitatum: accedunt notitia urbis Constantinopolitanae et Latercula Prouinciarum*, Frankfurt/Main: Minerva Verlag GMBH (orig. 1876).

Nov. Iust. = *Novellae Iustiniani*: ed. R. Schoell and G. Kroll (1954), *Corpus Iuris Civilis, vol.* IV. *Novellae*, Berlin: Weidmann.

Nov. Th. = *Novellae Theodosii*: ed. Th. Mommsen and P.M. Meyer ([4]1971), *Theodosiani libri* XVI *cum constitutionibus Sirmondianis et leges Novellae ad Theodosianum pertinentes*, Berlin, Weidmann (orig. 1902), II, 1–178; transl. Cl. Pharr (1952), *The Theodosian Code and Novels and the Sirmondian Constitutions*, Princeton: Princeton University Press, 487–514.

Sherk, Robert K. (1969), *Roman Documents from the Greek East. Senatus Consulta and Epistulae to the Age of Augustus*, Baltimore/Maryland: The Johns Hopkins Press.

Bibliography

Abel, Ernest L. (1968), "Were the Jews Banished from Rome in 19 A.D.?," *Revue des Études Juives* 127: 383–386.

Alföldi, Mária R. *et alii* (1957), *Intercisa II (Dunapentele). Geschichte der Stadt in der Römerzeit*, Budapest: Akadémiai Kiadó (2 vols.).

Alston, Richard (1995), *Soldier and Society in Roman Egypt*, London/New York: Routledge.

Alston, Richard (2002), "The Role of the Military in the Roman Revolution," *Aquila Legionis* 3: 7–42.

Ameling, Walter (2009), "The Epigraphic Habit and the Jewish Diasporas of Asia Minor and Syria," in H.M. Cotton, R.G. Hoyland, J.J. Price, and D.J. Wasserstein (eds.), *From Hellenism to Islam. Cultural and Linguistic Change in the Roman Near East*, Cambridge: Cambridge University Press, 203–234.

Anagnostou-Canas, Barbara (1989–1990), "Rapports de dépendance coloniale dans l'Egypte ptolemaique, I. L'appareil militaire," *Bullettino dell'Istituto di Diritto Romano "Vittorio Scialoja"* 31–32: 151–236.

Anagnostou-Canas, Barbara (1994), "La colonisation du sol dans l'Égypte ptolémaïque," in S. Allam (ed.), *Grund und Boden in Altägypten. Akten des internationalen Symposions Tübingen 18–20. Juni 1990*, Tübingen: Mohr Siebeck, 355–374.

Applebaum, Shim'on (1970), "Three Roman Soldiers of Probably Jewish Origin," in M. Rozelaar and B. Shimrom (eds.), *Commentationes ad antiquitatem classicam pertinentes in memoriam viri doctissimi Benzionis Katz*, Tel Aviv: Tel Aviv University Press, 3–6 [Hebrew].

Applebaum, Shim'on (1970), "Three Roman Soldiers of Probably Jewish Origin," in M. Rozelaar and B. Shimrom (eds.), (1971), "Jews and Service in the Roman Army," in *Roman Frontier Studies* 1967. *The Proceedings of the Seventh International Congress Held at Tel Aviv*, Tel Aviv: Tel Aviv University, 181–184.

Applebaum, Shim'on (1974), "The Legal Status of the Jews Communities in the Diaspora," in S. Safrai and M. Stern (eds.), *The Jewish People in the First Century. Historical Geography, Political History, Social, Cultural and Religious Life and Institutions*, Assen / Amsterdam: Van Gorcum, I, 420–463.

Applebaum, Shim'on (1979), *Jews and Greeks in Ancient Cyrene*, Leiden: Brill.

Applebaum, Shim'on (1989a), "Romanization and Indigenism in Judaea," in *Idem*, *Judaea in Hellenistic and Roman Times. Historical and Archaeological Essays*, Leiden: Brill, 155–165.

Applebaum, Shim'on (1989b), "The Troopers of Zamaris," in *Idem*, *Judaea in Hellenistic and Roman Times. Historical and Archaeological Essays*, Leiden: Brill, 47–65.

Askowith, Dora (1915), *The Toleration and Persecution of the Jews in the Roman Empire*,

1. The Toleration of the Jews under Julius Caesar and Augustus, New York: Columbia University.

Aster, Shawn Zelig (2019), "Sargon in Samaria—Unusual Formulations in the Royal Inscriptions and Their Value for Historical Reconstruction," *Journal of the American Oriental Society* 139 (3): 591–610.

Avery-Peck, Alan J. and Jacob Neusner (eds.), *Judaism in Late Antiquity, Part 3. Where We Stand: Issues and Debates in Ancient Judaism*, Leiden: Brill.

Avi-Yonah, Michael (1966), "The Epitaph of T. Mucius Clemens," *Israel Exploration Journal* 16: 258–264.

Avi-Yonah, Michael (1984), *The Jews under Roman and Byzantine Rule. A Political History of Palestine from the Bar Kokhba War to the Arab Conquest*, Jerusalem: The Magness Press/The Hebrew University (orig. Berlin: Walter de Gruyter, 1962).

Ayaso, José Ramón (1990), Iudaea Capta. *La Palestina romana entre las dos guerras judías (70–132 d. C.)*, Estella (Navarra): Verbo Divino.

Bacchiocchi, Samuele (1977), *From Sabbath to Sunday. A Historical Investigation of the Rise of Sunday Observance in Early Christianity*, Rome: The Pontifical Gregorian University.

Bagnall, Roger S. (1996), *Egypt in Late Antiquity*, Princeton: Princeton University Press (orig. 1993).

Bagnall, Roger S. and Raffaella Cribiore (1996), *Women's Letters from Ancient Egypt, 300 BC–AD 800*, Ann Arbor: The University of Michigan Press.

Baltrusch, Ernst (2002), *Die Juden und das römische Reich. Geschichte einer konfliktreichen Beziehung*, Darmstadt: Wissenschaftliche Buchgesellschaft.

Bar-Kochva, Bezalel (1979), *The Seleucid Army. Organization and Tactics in the Great Campaigns*, Cambridge: Cambridge University Press (orig. 1976).

Bar-Kochva, Bezalel (1989), *Judas Maccabaeus: The Jewish Struggle against the Seleucids*, Cambridge: Cambridge University Press.

Bar-Kochva, Bezalel (1996), *Pseudo Hecataeus, "On the Jews": Legitimizing the Jewish Diaspora*, Berkeley/Los Angeles: University of California Press.

Bar-Kochva, Bezalel (2010), *The Image of the Jews in Greek Literature. The Hellenistic Period*, Berkeley/Los Angeles: University of California Press.

Barclay, John M.G. (1996a), *Jews in the Mediterranean Diaspora from Alexander to Trajan (323 BCE–117 CE)*, Berkeley/Los Angeles/London: California University Press.

Barclay, John M.G. (1996b), "The Jews of the Diaspora," in J. Barclay and J. Sweet (eds.), *Early Christian Thought in Its Jewish Context*, Cambridge: Cambridge University Press, 27–40.

Barclay, John M.G. (1998), "Who Was Considered an Apostate in the Jewish Diaspora?" in G.N. Stanton and G.G. Stroumsa (eds.), *Tolerance and Intolerance in Early Judaism and Christianity*, Cambridge: Cambridge University Press, 80–95.

Barclay, John M.G. (2002), "Using and Refusing. Jewish Identity Strategies under the

Hegemony of Hellenism," in M. Konradt and U. Steinert (eds.), *Ethos und Identität. Einheit und Vielfalt des Judentums in hellenistisch-römischer Zeit*, Paderborn/Múnchen/Wien/Zürich: Ferdinand Schöningh, 13–25.

Barnes, Timothy D. (1993), *Athanasius and Constantius. Theology and Politics in the Constantinian Empire*, Cambridge (MA)/London: Harvard University Press.

Barnett, Paul W. (1974–1975), "Under Tiberius All Was Quiet," *New Testament Studies* 21: 564–571.

Baronowski, Donald Walter (1982), *Treaties of Military Alliance between Rome and Hellenistic States in the Last Three Centuries B.C.*, Toronto: University of Toronto (Ph. D. Dissertation).

Batsch, Christophe (2005), *La guerre et les rites de guerre dans le judaïsme du deuxième Temple*, Leiden/Boston: Brill.

Batsch, Christophe (2006), "Temps de la guerre et respect du sabbat dans Judith," in C. Grappe and J.-C. Ingelaere (eds.), *Le Temps et les Temps dans les littératures juives et chrétiennes au tournant de notre ère*, Leiden/Boston: Brill, 125–135.

Barkóczi, László (1964), "Population of Pannonia from Marcus Aurelius to Diocletian," *Acta Archaeologica Academiae Scientiarum Hungaricae* 16: 257–356.

Barkóczi, László *et alii* (1954), *Intercisa I (Dunapentele-Sztálinváros). Geschichte der Stadt in der Römerzeit*, Budapest: Akadémiai Kiadó (2 vols.).

Baudry, Gérard-Henry (2009), *Les symboles du christianisme ancien (Ier–VIIe siècle)*, Paris: Éditions du Cerf.

Belayche, Nicole (1999), "Dimenticare … Gerusalemme. Les paganismes à *Aelia Capitolina* du IIe au IVe siècle de notre ère," *Revue des Études Juives* 158: 287–348.

Belayche, Nicole (2003), "Tychè et la Tychè dans les cités de la Palestine romaine," *Syria* 80: 111–138.

Bellemore, Jane (1999), "Josephus, Pompey, and the Jews," *Historia. Zeitschrift für Alte Geschichte* 48: 94–118.

Benjamin, Lya, Mihai Spielmann, and S. Stanciu (1990), *Izvoare și mărturii referitoare la evreii din România, vol. 2, Part 2*, Bucharest: Hasefer.

Bermejo Rubio, Fernando (2020), *Los judíos en la Antigüedad. Desde el exilio en Babilonia hasta la irrupción del islam*, Madrid: Síntesis.

Bernand, André and Masson, Olivier (1957), "Les inscriptions grecques d'Abou-Simbel," *Revue des Études Grecques* 70 (329–330): 1–46.

Berthelot, Katell (2019), "Judaism as 'Citizenship' and the Question of the Impact of Rome," in K. Berthelot and J. Price (eds.), *In the Crucible of Empire. The Impact of Roman Citizenship upon Greeks, Jews and Christians*, Leuven/Paris/Bristol: Peeters, 107–129.

Bertolini, Dario (1874), "Iul. Concordia col. e la necropoli cristiana sopraterra recentemente scopertavi," *Archivio Veneto* 29: 276–300.

Bertolini, Dario and Wilhelm Henzen (1874), "Scavi concordiese," *Bullettino dell'Istituto di Corrispondenza Archeologica* Gennaio-Febbraio: 18–47.

Beševliev, Veselin (1967), "Tavernen als Kastellnamen bei Prokop," *Linguistique Balkanique* 12: 101–111.

Blau, Ludwig (1924), "Early Christian Epigraphy from the Jewish Point of View," *Hebrew Union College Annual* 1: 221–237.

Blau, Ludwig (1926), "Early Christian Archaeology from the Jewish Point of View," *Hebrew Union College Annual* 3: 157–214.

Blidstein, Gerald J. (1991), "The Import of Early Rabbinic Writings for an Understanding of Judaism in the Hellenistic-Roman Period," in Sh. Talmon (ed.), *Jewish Civilization in the Hellenistic-Roman Period*, Sheffield: Sheffield Academic Press, 64–72.

Block, René S. (2002), *Antike Vorstellungen vom Judentum: der Judenexkurs des Tacitus im Rahmen der griechisch-römischen Ethnographie*, Stuttgart Franz Steiner.

Blumenkranz, Bernhard (1960), *Juifs et Chrétiens dans le monde occidental, 430–1096*, Paris/La Haye: Mouton & Co.

Blumenkranz, Bernhard (1963), *Les auteurs chétiens latins du moyen âge sur les juifs et le judaïsme*, Paris/La Haye: Mouton & Co.

Bowersock, Glenn W. (1983), *Roman Arabia*, Cambridge (MA)/London: Harvard University Press.

Braun, Thomas (1998), "The Jews in the Late Roman Empire," *Scripta Classica Israelica* 17: 142–171.

Bringmann, Klaus (1983), *Hellenistische Reform und Religionsverfolgung in Judäa. Eine Untersuchung zur jüdisch-hellenistischen Geschichte (175–163 v. Chr.)*, Göttingen: Vandenhoeck & Ruprecht.

Broughton, Thomas Robert Shannon (2019), "The Roman Army," in Chr. B. Zeichmann (ed.), *Essential Essays for the Study of the Military in First-Century Palestine: Soldiers and the New Testament Context*, Eugene (OR): Pickwick Publications, 1–25.

Brunt, Peter Astbury (1974), "Conscription and Volunteering in the Roman Army," *Scripta Israelica Classica* 1: 90–115 (= *Idem* 1990, *Roman Imperial Themes*, Oxford: Oxford University Press, 189–214).

Buonopane, Alfredo (1994), "Un diploma militare ritrovato: autopsia di CIL, XVI, 8," *Epigraphica* 56: 216–223.

Burr, Victor Viktor (1972), "Rom und Judäa im 1. Jahrhundert v. Chr. (Pompeius und die Juden)," in H. Temporini (ed.), *Aufstieg und Niedergang der römischen Welt, II. Principat, 1. Religio*, Berlin/New York: Walter de Gruyter, 875–886.

Burr, Victor Viktor (2020), *Tiberio Julio Alejandro. Un aristócrata judío al servicio de Roma* (transl. and ed. by F. Bermejo Rubio), Madrid / Salamanca: Signifer Libros (orig. Bonn: Rudolf Habelt, 1955).

Burtchaell, James Tunstead (2004), *From Synagogue to Church. Public Services and Offices in the Earliest Christian Communities*, Cambridge: Cambridge University Press (orig. 1992).

Cameron, Duncan (2018), "Towards a Re-Dating of Pompey's Conquest of Jerusalem," *Journal of Jewish Studies* 69 (2), 225–247.

Campbell, Brian (1984), *The Emperor and the Roman Army, 31 BC–AD 235*, Clarendon Press, Oxford.

Campbell, Brian (2002), *War and Society in Imperial Rome, 31 BC–AD 284*, London/New York: Routledge.

Canfora, Luciano (2012), "Cesare e gli Ebrei," in G. Urso (ed.), *Iudaea socia, Iudaea capta. Atti del Convegno Internazionale Cividale del Friuli (22–24 settembre 2011)*, Pisa: Edizioni ETS, 97–105.

Canfora, Luciano (2012), "Cesare e gli Ebrei," in G. Urso (ed.) (2021), *Il tesoro degli Ebrei. Roma e Gerusalemme*, Bari/Roma: Gaiius. Laterza & Figli, 2021.

Cappelletti, Silvia (2006), *The Jewish Community of Rome. From the Second Century B.C. to the Third Century C.E.*, Leiden/Boston: Brill.

Capponi, Livia (2007), *Il Tempio di Leontopoli in Egitto: identità politica e religiosa dei giudei di Onia (c. 150 a.C.–73 d.C.)*, Pisa: ETS.

Carandini, Andrea, Andreina Ricci, and Mariette De Vos (1982), *Filosofiana, la villa di Piazza Armerina: immagine di un aristocratico romano al tempo di Constantino*, Palermo: S.F. Flaccovio.

Carletti, Carlo (2008), *Epigrafia dei cristiani in Occidente dal III al VII secolo. Ideologia e prassi*, Bari: Edipuglia.

Castelli, Silvia (2009), "Gli Ebrei espulsi da Roma e inviati in Sardegna da Tiberio nel 19 E.V. nelle fonti storiche di età romana," *Materia giudaica* 14 (1–2): 67–80.

Castritius, Helmut (2003a), "Waffenbesitz und Militärdienst der Juden von der römischen Antike bis zur Aufklärung. Zur Einführung," *Aschkenas. Zeitschrift für Geschichte und Kultur der Juden* 13: 1–2.

Castritius, Helmut (2003b), "Militärdienst und Wehrhaftigkeit der Juden in der Spätantike," *Aschkenas. Zeitschrift für Geschichte und Kultur der Juden* 13: 3–12.

Chastagnol, André (1977), "L'impôt payé par les soldats au IVe siècle," in *Armées et fiscalité dans le monde antique*, Paris: Éditions du Centre National de la Recherche Scientifique, 279–301.

Cheesman, George Leonard (1914), *The auxilia of the Roman Imperial Army*, Oxford: The Clarendon Press.

Christol, Michel (2006), "*Vir centenarius*," *Zeitschrift für Papyrologie und Epigraphik* 158: 243–250.

Chuvin, Pierre (²1991), *Chronique des derniers païens. La disparition du paganisme dans l'Empire romain, du règne de Constantin à celui de Justinien*, Paris: Les Belles Lettres/Fayard.

Cimma, Maria Rosa (1976), *Reges Socii et Amici Populi Romani*, Milano: Giuffrè.

Clarysse, Willy (1994), "Jews in Trikomia," in A. Bülow-Jacobsen (ed.), *Proceedings of the 20th International Congress of Papyrologists*, Copenhagen: Museum Tusculanum Press, 193–203.

Clarysse, Willy, Sofie Remijsen and Mark Depauw (2010), "Observing the Sabbath in the Roman Empire: a Case Study," *Scripta Classica Israelica* 29: 51–57.

Clauss, Manfred (1980), *Der magister officiorum in der Spätantike (4.–6. Jahrhundert). Das Amt und sein Einfluß auf die kaiserliche Politik*, München: C.H. Beck'sche.

Cohen, Getzel (1972), "The Hellenistic Military Colony: A Herodian Example," *Transactions and Proceedings of the American Philological Association* 103: 83–95.

Colafemmina, Cesare (1981), "Saggio di scavo in località 'Collina della Maddalena' a Venosa. Relazione preliminare," *Vetera Christianorum*, 18: 443–451.

Colafemmina, Cesare (2000), "Hebrew Inscriptions of the Early Medieval Period in Southern Italy," in B. Garvin and B. Cooperman (eds.), *The Jews of Italy. Memory and Identity*, Bethesda: University Press of Maryland, 65–81.

Collar, Anna (2013), *Religious Networks in the Roman Empire. The Spread of New Ideas*, Cambridge: Cambridge University Press.

Collins, John J. (2000), *Between Athens and Jerusalem. Jewish Identity in the Hellenistic Diaspora*, Grand Rapids (MI)/Cambridge: William B. Eermans Publishing/Dove Booksellers (2nd ed.).

Collins, John J. (2005), "The Jewish World and the Coming of Rome," in *Idem*, *Jewish Cult and Hellenistic Culture. Essays on the Jewish Encounter with Hellenism and Roman Rule*, Leiden/Boston: Brill, 202–215.

Colorni, Vittore (1956), *Gli Ebrei nel sistema del diritto comune fino alla prima emancipazione*, Milano: Giuffrè.

Cooley, Alison and Melvin George Lowe Cooley (^{2}2014), *Pompeii and Herculaneum: A Sourcebook*, London/New York: Routledge (orig. 2004).

Corbelli, Judith A. (2006), *The Art of Death in Graeco-Roman Egypt*, Pinces Risborough, UK: Shire.

Cornthwaite, Christopher J. (2017), "Wayward Jews, God-fearing Gentiles, or Curious Pagans? Jewish Normativity and the Sambathions," *Journal for the Study of Judaism* 48: 1–21.

Costa, José (2019), "Réflexions sur le judaïsme non-rabbinique, la rabbinisation et le judaïsme synagogal," *Judaïsme ancien-Ancient Judaism* 7: 187–241.

Cotton, Hannah M. (1995), "The Archive of Salome Komaïse Daughter of Levi: Another Archive from the 'Cave of Letters'," *Zeitschrift für Papyrologie und Epigraphik* 105: 171–208.

Cotton, Hannah M. (1998), "The Rabbis and the Documents," in M. Goodman (ed.), *Jews in a Graeco-Roman World*, Oxford: Clarendon Press, 167–179.

Cowey, James M.S. and Klaus Maresch, (2001) *Urkunden des Politeuma der Juden von Herakleopolis (144/3–133/2 v. Chr.)*, Wiesbaden: Westdeutscher.

Cracco Ruggini, Lellia (2003), "A proposito di Dura Europos: L'acculturazione ebraica nel tardo ellenismo," *Mediterraneo Antico. Economie, Società, Culture* 6 (1), 2003: 661–682.

Cresci Marrone, Giovannella (2001), "Lo stanziamento militare, la fabbrica di frecce e la comunità di commercianti orientali nella Concordia tardo antica," in P. Croce da Villa and E. di Filippo Balestrazzi (eds.), *Concordia Sagittaria. Tremila anni di storia*, Concordia: Comune di Concordia Saggitaria, 245–249.

Currant, John (2015), "Jews," in Y. Le Bohec *et alii* (eds.), *The Encyclopedia of the Roman Army*, Oxford: Wiley Blackwell, II, 559.

Cuscito, Giuseppe (2009), "Cimiteri e santuari paleocristiani dell'Alto Adriatico da Concordia a Trieste," in *Idem*, Signaculum fidei. *L'ambiente cristiano delle origini nell'Alto Adriatico: aspetti e problemi*, Trieste: Editreg, 149–176 [orig. *Idem* (2004), "Santuari e cimiteri sul territorio di Aquileia nella tarda antichità," *Antichità Altoadriatiche* 57: 191–242].

Czajkowski, Kimberley (2015), "Jewish Attitudes towards the Imperial Cult," *Scripta Classica Israelica* 34: 181–194.

Czajkowski, Kimberley (2017), *Localized Law. The Babatha and Salome Komaïse Archives*, Oxford: Oxford University Press.

Dalley, Stephanie (1985), "Foreign Chariotry and Cavalry in the Armies of Tiglath-Pileser III and Sargon II," *Iraq* 47: 31–48.

Darmesteter, Arsène (1880), "Notes épigraphiques touchant quelques points de l'histoire des Juifs sous l'empire romain," *Revue des Études Juives*, 1: 32–55.

Davenport, Caillan (2012), "Soldiers and Equestrian Rank in the Third Century AD," *Papers of the British School at Rome* 80: 89–123.

De Bonfils, Giovanni (1998), Omnes … ad implenda munia teneantur. *Ebrei curie e prefetture fra IV e V secolo*, Bari: Cacucci Editore.

De Bonfils, Giovanni (2005), *Gli Ebrei dell'Impero di Roma*, Bari: Cacucci Editore.

de Falco, Diana Joyce (2017), "Epitaffio di Lucius Aiacius Dama," in A. Foa, G. Lacerenza, and D. Jalla (eds.), *Ebrei, una storia italiana. I primi mille anni*, Milano: Electa, 267.

de Lange, Nicholas (1978), "Jewish Attitudes to the Roman Empire," in P.D.A. Garnsey and C.R. Whittaker (eds.), *Imperialism in the Ancient World*, Cambridge: Cambridge University Press, 255–281 and 354–357.

de Lange, Nicholas (2005), "Jews in the Age of Justinian," in M. Maas (ed.), *The Cambridge Companion to the Age of Justinian*, Cambridge: Cambridge University Press, 401–426.

de Lange, Nicholas (2017), "La imagen de Roma en las fuentes rabínicas," in G. Bravo and R. González Salinero (eds.), *Ideología y religión en el mundo romano*, Madrid/Salamanca: Signifer Libros, 21–38.

Dello Russo, Jessica (2010), "The Monteverde Jewish Catacombs on the via Portuense," *Roma Subterranea Judaica* 4: 1–37.

Dello Russo, Jessica (2014), "Recent Studies on the Jewish Catacombs of Monteverde in Rome: A Review Article," *Sefer yuḥasin. Review for the History of the Jews in South Italy* 3: 9–63.

Delmaire, Roland (1995), *Les institutions du Bas-Empire romain, de Constantin à Justinien, I. Les institutions civiles palatines*, Paris: Les Éditions du Cerf.

Demougeot, Émilienne (1960), "L'empereur Honorius et la politique antijuive," in *Hommages à Léon Herrmann*, Bruxelles/Berchem: Latomus, 277–291.

Demougeot, Émilienne (1961), "'Paganus', Mithra et Tertullien," in F.L. Cross (ed.), *Studia Patristica, III*, Berlin: Akademie Verlag, pp. 354–365.

Devijver, Hubert (1976–1977), *Prosopographia militiarum equestrium que fuerunt ab Augusto ad Gallienum*, Leuven: Universitaire Pers (2 vols.).

Dezsö, Tamás (2012), *The Assyrian Army*, Budapest: Eötvös University Press (2 vols.).

Di Paola, Lucietta (2005), *Per la storia degli "occhi del re". I servizi ispettivi nella Tarda Antichità*, Messina: Università degli Studi di Messina.

Di Paola, Lucietta (2009), "I *curiosi* in eta tardoantica: riflessionii in margine al titolo VI, 29 del Teodosiano," in S. Crogiez-Pétrequin and P. Jaillette (eds.), *Le Code Théodosien. Diversité des approches et nouvelles perspectives*, Rome: École Française de Rome, 119–141.

Di Paola, Lucietta (2013), "*Agentes in rebus, curiosi, basileos, ophthalmoi, peuthenes* e *magistriani* all'opera nel controllo dei corpi e delle menti," in F. de Marini Avanzo (ed.), *Atti dell'Accademia Romanistica Costantiniana, XIX. Organizzare Sorvegliare Punire: Il controllo dei corpi e delle menti nel diritto della tarda antichità*, Roma: Aracne, 297–312.

di Stefano Manzella, Ivan (1987), *Mestiere di epigrafista. Guida alla schedatura del materiale epigrafico lapideo*, Roma: Quasar.

di Stefano Manzella, Ivan (1989), "*L. Maecius archon, centurio alti ordinis*. Nota critica su *CIL*, VI, 39084 = *CII*, I, 470," *Zeitschrift für Papyrologie und Epigraphik* 77: 103–112.

Dirven, Lucinda (1999), *The Palmyrenes of Dura-Europos: A Study of Religious Interaction in Roman Syria*, Leiden: Brill.

Dirven, Lucinda (2007), "The Julius Terentius fresco and the Roman Imperial Cult," *Mediterraneo antico* 10 (1–2): 115–127.

Dobson, Brian (1974), "The Significance of the centurion and 'primipilaris' in the Roman Army and Administration," in H. Temporini (ed.), *Aufstieg und Niedergang der römischen Welt*, II.1, Berlin/New York: Walter de Gruyter, 392–434.

Doering, Lutz (1997), "The Concept of the Sabbath in the Book of Jubilees", in M. Albani, J. Frey and A. Lange (eds.), *Studies in the Book of Jubilees*, Tübingen: Mohr Siebeck, 179–205.

Drijvers, Han (1992), "Syrian Christianity and Judaism," in J. Lieu, J. North and T. Rajak (eds.), *The Jews among Pagans and Christians in the Roman Empire*, London/New York: Routledge, 124–146.

Eck, Werner (2010), "A Second Constitution for the Auxiliary Troops in Judaea in 86 AD," *Scripta Classica Israelica* 29: 21–31.

Eck, Werner and Andreas Pangerl (2015), "Eine zweite Kopie der Konstitution für die Truppen Syriens vom 19. März 144 und ein Diplom für die equites singulares vom selben Datum," *Zeitschrift für Papyrologie und Epigraphik* 193: 253–260.

Eck, Werner and Dirk Koßmann (2015), "Ein Legionssoldat in einer Inschrift aus Nazareth und die Präsenz der *legio IV Flavia Felix* in den östlichen Provinzen," *Scripta Judaica Cracoviensia* 13: 31–40.

Eckhardt, Benedikt (ed.) (2019), *Private Associations and Jewish Communities in the Hellenistic and Roman Cities*, Leiden: Brill.

Empereur, Jean-Yves (1995), *A Short Guide to the Catacombs of Kom El Shoqafa*, Alexandria, Alexandria: Serapis.

Empereur, Jean-Yves (2018), "New Data concerning the Foundation of Alexandria," in Chr. S. Zerefos and M.V. Verdinoyannis (eds.), *Hellenistic Alexandria. Celebrating 24 Centuries*, Oxford: Archaeopress Publishing, 3–12.

Etienne, Stéphane (2000), "Réflexion sur l'apostasie de Tibérius Julius Alexander", *The Studia Philonica Annual* 12: 122–142.

Eshel, Esther and Stone, Michael E. (2013), "Judaism in Palestine in the Hellenistic-Roman Periods," in M.R. Salzman and W. Adler (eds.), *The Cambridge History of Religions in the Ancient World, II. From the Hellenistic Age to Late Antiquity*, Cambridge: Cambridge University Press, 87–115.

Faber, Eike (2013), "How Foreign Soldiers Became more Important than Roman Citizens. The Development of Differences in Status between Julius Caesar and Theodosius I," in D. Álvarez Jiménez, R. Sanz Serrano and D. Hernández de la Fuente (eds.), *El espejismo del bárbaro. Ciudadanos y extranjeros al final de la Antigüedad*, Castelló de la Plana: Universitat Jaume I, 115–130.

Fasola, Umberto Maria (1976), "Le due catacombe ebraiche di Villa Torlonia," *Rivista di Archeologia Cristiana* 52: 7–62.

Feldman, Louis H. (1960), "The Orthodoxy of the Jews in Hellenistic Egypt," *Jewish Social Studies* 22: 212–237.

Feldman, Louis H. (1993), *Jew and Gentile in the Ancient World. Attitudes and Interactions from Alexander to Justinian*, Princeton (NJ): Princeton University Press.

Fernández Ubiña, José (2000), *Cristianos y militares. La Iglesia antigua ante el ejército y la guerra*, Granada: Universidad de Granada.

Ferrua, Antonio (1936), "Sulla tomba dei Cristiani e su quella degli Ebrei," *La Civiltà Cattolica* 87 (4): 298–311.

Figueras, Pau (2000–2002), "Motivos paganos en mosaicos cristianos y judíos de Oriente: problemática e interpretación," *Espacio, Tiempo y Forma, II. Historia Antigua* 13: 261–296, 297–320; 14: 129–169; 15: 43–86, 241–274.

Firpo, Giulio (2009), "La guerra giudaica e l'ascesa di Vespasiano," in F. Coarelli (ed.), *Divus Vespasianus. II bimillenario dei Flavi*, Milano: Electa, 42–45.

Fischer-Bovet, Christelle (2014), *Army and Society in Ptolemaic Egypt*, Cambridge: Cambridge University Press.

Fitz, Jenö (1972), *Les Syriens à Intercisa*, Bruxelles: Latomus.

Foot Moore, George (1948–1954), *Judaism in the First Centuries of the Christian Era. The Age of the Tannaim*, Cambridge (Mass.): Harvard University Press.

France, Jérôme and Jocelyne Nelis-Clément (2014), "La *statio*. Archéologie d'un lieu de pouvoir dans l'empire romain," in J. France and J. Nelis-Clément (eds.), *La* statio. *Archéologie d'un lieu de pouvoir dans l'empire romain*, Bordeaux/Paris: Ausonius/De Boccard (Scripta Antiqua, 66), 11–15.

Frankel, Ellen and Betsy Platkin Teutsch (1995), *Encyclopedia of Jewish Symbols*, Northvale (NJ): Janson Aronson.

Friedheim, Emmanuel (2015), "Le Milieu rabbinique face aux diversités sociales et religieuses en Palestine romaine du Ier au IVe siècle de notre ère," in M.e-A. Vannier (ed.), *Judaïsme et christianisme chez les Pères*, Turnhout: Brepols, 147–164.

Friedheim, Emmanuel (2020), "Le judéo-polythéisme en Galilée au temps de la Mišna et du Talmud-sources complémentaires," *Revues de Études Juives* 179(3–4): 291–313.

Fülep, Ferenc (1954), "Epigraphie," in L. Barkóczi *et alii*, *Intercisa I (Dunapentele-Sztálinváros). Geschichte der Stadt in der Römerzeit*, Budapest: Akadémiai Kiadó, 232–273.

Gabba, Emilio (1977), "The Origins of the Professional Army at Rome: The 'proletarii' and Marius' Reform," in *Idem*, *Republican Rome, the Army and the Allies* (transl. P.J. Cuff), Oxford: Basil Blackwell, 1–19 (= *Athenaeum* 27, 1949, 173–209).

Gabba, Emilio (1999), "The Social, Economic and Political History of Palestine 63 BCE–CE 70," in W. Horbury, W.D. Davies, and J. Sturdy (eds.), *The Cambridge History of Judaism, III. The Early Roman Period*, Cambridge: Cambridge University Press, 94–167.

Gafni, Chanan (2016), "Orthodoxy and Talmudic Criticism? On Misleading Attributions in the Talmud," *Zutot. Perspectives on Jewish Culture* 13: 70–80.

Gaudemet, Jean (1988), "La condition juridique des Juifs dans les trois premiers siècles de l'Empire," *Augustinianum* 28: 339–359.

Gauthier, François (2020), "The Transformation of the Roman Army in the Last Decades of the Republic," in J. Armstrong and M.P. Fronda (eds.), *Romans at War. Soldiers, Citizens, and Society in the Roman Republic*, London/New York: Routledge, 283–296.

Gichon, Mordechai (2009), "The Manifestations of the Religious Cults of the Roman Army and Administration, in Conflontation with the Observance of Its Jewish Population," in C. Wolff and Y. Le Bohec (eds.), *L'armée romaine et la religion sous le Haut-Empire romain*, Lyon: De Boccard, 149–158.

Gichon, Mordechai and Michaela Vitale (1991), "Arrow-heads from Ḥorvat 'Eqed," *Israel Exploration Journal* 41(4): 242–257.

Ginestet, Pierre (1991), *Les organisations de la jeunesse dans l'Occident Romain*, Bruxelles: Latomus.

Ginzberg, Louis (1909–1938), *The Legends of the Jews* (transl. H. Szold *et alii*), Philadelphia: Jewish Publication Society of America (7 vols.).

Ginzberg, Louis (1962), *On Jewish Law and Lore*, Cleveland/New York: The World Publishing Company.

Glay, Marcel (1972), "Le commandement des *cohortes voluntariorum* de l'armée romaine," *Ancient Society* 3: 209–222.

Goldenberg, Robert (1979), "The Jewish Sabbath in the Roman World up to the Time of Constantine the Great," in W. Haase (ed.), *Aufstieg und Niedergang der römischen Welt, II. Principat, 1. Religio*, Berlin/New York: Walter de Gruyter, 414–447.

Goldstein, Jonathan A. (1989), "The Hasmonean Revolt and the Hasmonean Dynasty," in W.D. Davies and L. Finkelstein (eds.), *The Cambridge History of Judaism, II. The Hellenistic Age*, Cambridge: Cambridge University Press, 292–351.

Goldsworthy, Adrian (1996), *The Roman Army at War, 100 B.C.–A.D. 200*, Oxford: The Clarendon Press.

González Salinero, Raúl (2000), *El antijudaísmo cristiano occidental (siglos IV y V)*, Madrid: Trotta.

González Salinero, Raúl (2003), "El servicio militar de los judíos en el ejército romano," *Aquila Legionis* 4: 45–91.

González Salinero, Raúl (2008), "Investigadores de la corrupción, corruptos: la degradación moral de los *agentes in rebus*," in G. Bravo and R. González Salinero (eds.), *La corrupción en el mundo romano*, Madrid/Salamanca: Signifer Libros, 191–207.

González Salinero, Raúl (2019a), "Indicios de la política favorable de Septimio Severo hacia los judíos," in J. Cabrero Piquero and P. González Serrano (eds.), *Estudios sobre el Mundo Antiguo dedicados a la Profesora Pilar Fernández Uriel*, Madrid/Salamanca: Signifer Libros, 237–250.

González Salinero, Raúl (2019b), "La *Altercatio: Ecclesiae et Synagogae*: de la retórica exegética a la realidad histórica," *Bollettino di Studi Latini* 49: 572–583.

Goodenough, Erwin R. (1953–1968), *Jewish Symbols in Greco-Roman Period*, New York: Pantheon Books (13 vols.).

Goodenough, Erwin R. (1988), *Jewish Symbols in the Greco-Roman Period* (Abridged Edition by J. Neusner), Princeton: Princeton University Press.

Goodman, Martin D. (1987), *The Ruling Class of Judaea. The Origins of the Jewish Revolt against Rome A.D. 66–70*, Cambridge: Cambridge University Press.

Goodman, Martin D. (2004), "Sadducees and Essenes after 70 CE," in M. Goodman (ed.), *Judaism in the Roman World*, Leiden: Brill, 153–162.

Goodman, Martin D. (2007), *Rome and Jerusalem. The Clash of Ancient Civilizations*, London: Allen Lane (= New York: Vintage, 2008).

Goodman, Martin D. and A.J. Holladay (1986), "Religious Scruples in Ancient Warfare," *The Classical Quarterly* 36 (1): 151–171.

Goodnick Westenholz, Joan (1994), *The Jewish Presence in Ancient Rome*, Jerusalem: Bible Lands Museum Jerusalem.

Grabbe, Lester L. (2008), *A History of the Jews and Judaism in the Second Period, 2.*

The Coming of the Greeks: The Early Hellenistic Period (*335–175 BCE*), Edinburgh: T&T Clark.

Gracey, Michael H. (1986), "The Armies of the Judaean Client Kings," in Ph. Freeman and D. Kennedy (eds.), *The Defence of the Roman and Byzantine East. Part 1*, Oxford: Bar International Series, 311–323.

Grainger, John D. (2013), *Roman Conquest: Egypt and Judaea*, Barsley (South Yorkshire): Pen & Sward Military.

Granerød, Gard (2019), "Canon and Archive: Yahwism in Elephantine and Āl-Yāḫūdu as a Challenge to the Canonical History of Judean Religion in the Persian Period," *Journal of Biblical Literature* 138 (2): 345–364.

Grant, Michael (1999), *The Jews in the Roman World*, London: Phoenix Giant, 1999 (orig. London: Weidenfeld & Nicolson, 1973).

Grayzel, Solomon (1968), "The Jews and Roman Law," *Jewish Quarterly Review* 59: 93–117.

Green, Deborah (2008), "Sweet Spices in the Tomb: An Initial Study on the Use of Perfume in Jewish Burials," in L. Brink and D. Green (eds.), *Commemorating the Dead: Texts and Artifacts in Context. Studies of Roman, Jewish, and Christian Burials*, Berlin/New York: Walter de Gruyter, 145–173.

Grotius, Hugo (2012) *On the Law of War and Peace* (*De jure belli ac pacis*) (transl. and annotated by St.C. Neff), Cambridge: Cambridge University Press (orig. Paris 1624; Amsterdam 1631^2).

Gruen, Erich S. (1984), *The Hellenistic World and the Coming of Rome*, Berkeley: University of California Press.

Gruen, Erich S. (2002), *Diaspora. Jews amidst Greeks and Romans*, Cambridge (Massachusetts) / London: Harvard University Press.

Gruen, Erich S. (2009), "Tacitus and the Defamation of the Jews," in J. Geiger, H. Cotton and G. Stiebel (eds.), *Israel's Land. Papers Presented to Israel Shatzman on His Jubilee*, Jerusalem: The Open University of Israel/Israel Exploration Society, 77–96 (= *Idem* (2016), *Constructs of Identity in Hellenistic Judaism. Essays on Early Jewish Literature and History*, Walter de Gruyter, Berlin/Boston, 265–280).

Gudea, Nicolae (1999–2000), "Evreii în provinciile dacice 106–275 p. Ch." *Ephemeris Napocensis* 9–10: 179–208.

Guichaoua, Michaël (2019), *Les unités de l'armée romaine tardive (360–423). Cohérence de la Notice des Dignités, et compréhension de la crise militaire sous Théodose et Honorius*, Wrocław: Amazon Fulfillment.

Günther, Linda-Marie (2007), "Herodes, Caesar (Augustus) und Caesarea," in L.-M. Günther (ed.), L.-M. Günther (ed.), *Herodes und Rom*, Stuttgart: Franz Steiner, 79–89.

Gwynn, David M. (2007), *The Eusebians. The Polemic of Athanasius of Alexandria and the Construction of the 'Arian Controversy'*, Oxford: Oxford University Press.

Gwynn, David M. (2012), *Athanasius of Alexandria. Bishop, Theologian, Ascetic, Father*, Oxford: Oxford University Press.

Hachlili, Rachel (2001), *The Menorah, the Ancient Seven-Armed Candelabrum. Origin, Form and Significance*, Leiden/Boston/Köln: Brill.

Hadas-Lebel, Mireille (2009), *Rome, la Judée et les Juifs*, Paris: A. et J. Picard.

Haensch, Rudolf (2014), "The Contributions of Inscriptions to our Knowledge of the Herodian Dynasty," *Scripta Classica Israelica* 33: 99–116.

Harland, Philip A. (2003), *Associations, Synagogues and Congregations. Claiming a Place in Ancient Mediterranean Society*, Minneapolis: Fortress Press.

Harmand, Jacques (1967), *L'armée et le soldat à Rome de 107 à 50 avant notre ère*, Paris: Picard.

Harrington, Hannah K. (2001), *Holiness. Rabbinic Judaism and the Graeco-Roman World*, London/New York: Routledge.

Hayes, Christine (2007), "The Other in Rabbinic Literature," in C. Fonrobert and M. Jaffe (eds.), *The Cambridge Companion to the Talmud and Rabbinic Literature*, Cambridge: Cambridge University Press, 243–269.

Haynes, Ian (2013), *Blood of the Provinces. The Auxilia and the Making of Provincial Society from Augustus to the Severans*, Oxford: Oxford University Press.

Hebblewhite, Mark (2017), *The Emperor and the Army in the Later Roman Empire, AD 235–395*, New York: Routledge.

Hegermann, Harald (1989), "The Diaspora in the Hellenistic Age," in W.D. Davies and L. Finkelstein (eds.), *The Cambridge History of Judaism, II. The Hellenistic Age*, Cambridge: Cambridge University Press, 115–166.

Heidel, W.A. (1920), "Why Were the Jews Banished from Italy in 19 A.D.?" *American Journal of Philology* 41: 38–47.

Helgeland, John (1978), "Roman Army Religion," in W. Haase (ed.), *Aufstieg und Niedergang der römischen Welt, II. Principat, 2*, Berlin/New York: Walter de Gruyter, 1470–1505.

Hengel, Martin (1989), "The Political and Social History of Palestine from Alexander to Antiochus III (333–187 B.C.E.)," in W.D. Davies and L. Finkelstein (eds.), *The Cambridge History of Judaism, II. The Hellenistic Age*, Cambridge: Cambridge University Press, 35–79.

Henten, Jan Willem van, and Friedrich Avemarie (2006), "Martirio e morte nobile nel giudaismo e nelle fonti dell'antico ebraismo ellenistico," *Cristianesimo nella Storia* 27: 31–65.

Herr, Moshe David (1960–1961), "The Problem of War on the Sabbath in the Second Temple and the Talmudic Period," *Tarbiẓ* 30: 242–256 and 341–356. [Hebrew, English summary, VII–IX].

Herz, Peter D. (2002), "*Agentes in rebus*," in H. Cancik and A. Schneider (eds.), *Brill's Encyclopaedia of the Ancient World. New Pauly, Antiquity*, Leiden: Brill, vol. 1, 325.

Hezser, Catherine (2011), *Jewish Travel in Antiquity*, Tübingen: Mohr Siebeck.

Hezser, Catherine (2016), "Mobility, Flexibility, and the Diasporization of Palestinian Judaism after 70 C.E.," in C. Cordoni and G. Langer (eds.), *"Let the Wise Listen and Add to Their Learning" (Prov 1:5). Festschrift for Günter Stemberger on the Occasion of His 75th Birthday*, Berlin/Boston: Walter de Gruyter, 197–216.

Hild, Joseph-Antoine (1884–1885), "Les Juifs à Rome devant l'opinion et dans la littérature," *Revue des Études Juives* 8: 1–37; 10: 18–59; 11: 161–194.

Hirschfeld, Amy K. (2008), "An Overview of the Intellectual History of Catacomb Archaeology," in L. Brink, O.P. and D. Green (eds.), *Commemorating the Death. Texts and Artifacts in Contexts. Studies of Roman, Jewish, and Christian Burials*, Berlin/New York: Walter de Gruyter, 11–38.

Hoffmann, Dietrich (1963), "Die spätrömischen Soldatengrabinschriften von Concordia," *Museum Helveticum*, 20 (1): 22–57.

Hoffmann, Dietrich (1969–1970), *Das spätrömische Bewegungsheer und die* Notitia Dignitatum, Düsseldorf: Rheinland-Verlag (2 vols.).

Holder, Paul A. (1980), *Studies in the Auxilia of the Roman Army from Augustus to Trajan*, Oxford: BAR (BARIS 70).

Honigman, Sylvie (2003) "*Politeumata* and Ethnicity in Ptolemaic and Roman Egypt," *Ancient Society* 23: 61–102.

Honigman, Sylvie (2016), "The Ptolemaic and Roman Definition of Social Categories and the Evolution of Judaean Communal Identity in Egypt," in Y. Furstenberg (ed.), *Jewish and Christian Communal Identities in the Roman World*, Leiden/Boston: Brill, 25–74.

Horbury, William (2014), *Jewish War under Trajan and Hadrian*, Cambridge: Cambridge University Press.

Hudson, Elizabeth Harriot (1882), *A History of the Jews in Rome, B.C. 160–A.D. 604*, London: Hodder & Stoughton.

Ilan, Tal (2008), *Lexicon of Jewish Names in Late Antiquity, III. The Western Diaspora 330 BCE–650 CE*, Tübingen: Mohr Siebeck (in collaboration with Thomas Ziem).

Ilan, Tal (2009), "The Torah of the Jews of Ancient Rome," *Jewish Studies Quarterly* 16: 363–395.

Impallomeni, Giambattista (1984), "Per una nuova ipotesi sul fondamento giuridico delle sanzioni sepolcrali alla luce dei ritrovamenti in Concordia Sagittaria," *Aquileia Nostra* 55, 121–136.

Instone-Brewer, David (2011), *Feasts and Sabbaths. Passover and Atonement*, Grand Rapids (Michigan)/Cambridge: William B. Eerdmans.

Isaac, Benjamin (1990), *The Limits of Empire. The Roman Army in the East*, Oxford: Clarendon Press.

Isaac, Benjamin (1991), "The Roman Army in Judaea: Policy Duties and Taxation,", in V.A. Maxfiel and M.J. Dobson (eds.), *Roman Frontier Studies 1989*, Exeter: University of Exeter, 458–461.

Isaac, Benjamin (1998), "Roman Colonies in Judaea: The Foundation of Aelia Capitolina," in *Idem, The Near East under Roman Rule. Selected Papers*, Leiden/New York/Köln: Brill, 87–111 (orig. in *Talanta* 12, 1980–1981: 31–54).

Jacobson, David M. (2001), "Three Roman Client Kings: Herod of Judaea, Archelaus of Cappadocia and Juba of Mauretania," *Palestine Exploration Quarterly* 133: 22–38.

Jacobson, David M. (2019), *Agrippa II. The Last of the Herods*, London/New York: Routledge.

James, Simon (2004), *The Excavations at Dura-Europos Conducted by Yale University and the French Academy of Inscriptions and Letters (1928 to 1937). Final Report VII. The Arms, Armour, and Other Military Equipment*, London: British Museum Press.

James, Simon (2019), *The Roman Military Base at Dura-Europos, Syria. An Archaeological Visualization*, Oxford: Oxford University Press.

Jeffers, J.S. (1998), "Jewish and Christian Families in First-Century Rome," in K.P. Donfried and P. Richardson (eds.), *Judaism and Christianity in First-Century Rome*, Grand Rapids, MI/Cambridge: William B. Eerdmans, 128–150.

Jiménez de Furundarena, Agustín (2018), "*Cohors I Ascalonitanorum sagittaria equitata*," *Aquila Legionis* 21: 81–105.

Johns, Alger F. (1963), "The Military Strategy of Sabbath Attacks on the Jews," *Vetus Testamentum* 13 (4): 482–486.

Johnston, David (2004), *Roman Law in Context*, Cambridge: Cambridge University Press.

Joisten-Pruschke, Anke (2008), *Das religiöse Leben der Juden von Elephantine in der Achämenidenzeit*, Wiesbaden: Harrassowitz.

Jones, Arnold Hugh Martin (1992), *The Roman Later Empire 284–602. A Social, Economic, and Administrative Survey*, The Johns Hopkins University Press, Baltimore, 1992 (orig. Oxford: Basil Blackwell, 1964, 2 vols.)

Jones, Arnold Hugh Martin, John Robert Martindale, and John Morris (1971), *The Prosopography of the Later Roman Empire, I. A.D. 260–395*, Cambridge: Cambridge University Press (*PLRE I*).

Jordán Montes, Juan Francisco (1997), "Los judíos en el reinado de Honorio (395–423 d.C.)," in R. Teja and C. Pérez (eds.), *Congreso internacional "La Hispania de Teodosio"*, Salamanca: Junta de Castilla y León/Universidad SEK, I, 121–134.

Juster, Jean (1914), *Les Juifs dans l'Empire romain. Leur condition juridique, économique et sociale*, Paris: Librairie Paul Geuthner, 2 vols.

Kaiser, Ted (2017), "Empire Community, and Culture on the Middle Euphrates. Durenes, Palmyrenes, Villagers, and Soldiers," *Bulletin of the Institute of Classical Studies* 60 (1): 63–95.

Kajanto, Iiro (1982), *The Latin Cognomina*, Roma: Giorgio Bretschneider (orig. Helsinki: Helsingfors, 1965).

Kaplan, Philip (2003), "Cross-Cultural Contacts among Mercenary Communities in Saite and Persian Egypt," *Mediterranean Historical Review* 18: 1–31.

Kasher, Aryeh (1978), "First Jewish Military Units in Ptolemaic Egypt," *Journal for the Study of Judaism*, 9 (1): 57–67.

Kasher, Aryeh (1985), *The Jews in Hellenistic and Roman Egypt. The Struggle for Equal Rights*, Tübingen: J.C.B. Mohr (Paul Siebeck).

Keenan, James G. (1973), "The Names Flavius and Aurelius as Status Designations in Later Roman Egypt," *Zeitschrift für Papyrologie und Epigraphik* 11, 33–63.

Kendall, Seth (2013), *The Struggle for Roman Citizenship. Romans, Allies, and the Wars of 91–77 BCE*, Piscataway, NJ: Gorgias Press.

Kennedy, David (2004), *The Roman Army in Jordan*, London: Council for British Research in the Levant (orig. 2000).

Keppie, Lawrence (1984), *The Making of the Roman Army*, Totowa: Barnes and Noble.

Kokkinos, Nikos (1990), "A Fresh Look at the *gentilicium* of Felix, Procurator of Judaea," *Latomus* 49: 1126–1241.

Kotansky, Roy (1994), *Greek Magical Amulets. The Inscribed Gold, Silver, Copper, and Bronze* Lamellae, *Part I. Published Texts of Known Provenance*, Wiesbaden: Springer Fachmedien.

Kovács, Péter (2004), "The Late Roman Epigraphy in Pannonia (260–582)," *Hungarian Polis Studies* 11: 185–195.

Kovács, Péter (2010), "Notes on the 'Jewish' Inscriptions in Pannonia," *Journal of Ancient Judaism* 1: 159–180.

Kraeling, Carl H. (1956), *The Synagogue. The Excavations at Dura-Europos. Final Report VIII. Part I*, New Haven: Yale University Press.

Kraemer, Ross Shepard (2020), *The Mediterranean Diaspora in Late Antiquity. What Christianity Cost the Jews*, Oxford: Oxford University Press.

Kraft, Konrad (1951), *Zur Rekrutierung der Alen und Kohorten an Rhein und Donau*, Berna: A. Francke.

Krause, Andre R. (2017), *Synagogues in the Works of Flavius Josephus: Rhetoric, Spatiality, and First-Century Jewish Institutions*, Leiden/Boston: Brill.

Krauss, Samuel (1922), *Synagogale Altertümer*, Berlin/Wien: Nachdruck der Ausgabe (reed. Hildesheim: George Olms, 1966).

Kromayer, Johannes and Georg Veith (1928), *Heerwesen und Kriegführung der Griechen und Römer*, München: C.H. Beck'sche.

Kubo, Sakae (1982), "The Sabbath in the Intertestamental Period," in K.A. Strand (ed.), *The Sabbath in Scripture and History*, Washington D.C.: Review and Herald Publishing Corporation, 57–69.

Kuhnen, Hans-Peter (1987), *Nordwest-Palästina in hellenistisch-römischer Zeit*, Weinheim: VCH Acta Humaniora.

Lafli, Ergün (2021), "Identity of the Jewish Community in Roman Aeolis: Remarks on Two Inscriptions from Izmir in Western Turkey," in L. Locatelli, É. Piguet, and S. Podestà (eds.), *Constructions identitaires en Asie Mineure* (*VIII*[e] *siècle avant J.-C.–III*[e] *siècle après J.-C.*), Paris: Presses Universitaires de Franche-Comté, 2021, 333–370.

Laham Cohen, Rodrigo (2019), "La historia lacrimógena, la diáspora antes de la diáspora y los peligros de estudiar exclusivamente a los judíos," *Sociedad y Religión* 29 (52), 127–147.

Laham Cohen, Rodrigo (2021), "Identidades judías en la Europa tardoantigua: Erwin Goodenough y sus ecos," *Sociedades Precapitalistas* 11: (e063) 1–26.

Lapin, Hayim (2012), *Rabbis as Romans. The Rabbinic Movement in Palestine, 100–400 C.E.*, Oxford: Oxford University Press.

Launey, Marcel (1987), *Recherches sur les armées hellénistiques* (ed. by Y. Garlan, Ph. Gauthier, and Cl. Orrieux), Paris: De Boccard. 2 vols. (orig. 1949–1950).

Laurenzi, Elsa (2011), *Le catacombe ebraiche. La comunità ebraica dell'antica Roma e le sue tradizioni funerarie*, Roma: Gangemi Editore.

Lavan, M. (2019a), "The Foundation of Empire? The Spread of Roman Citizenship from the Fourth Century BCE to the Third Century CE," in K. Berthelot and J. Price (eds.), *In the Crucible of Empire. The Impact of Roman Citizenship upon Greeks, Jews and Christians*, Leuven/ Paris/Bristol: Peeters, 21–54.

Lavan, M. (2019b), "The Roman Army and the Spread of Roman Citizenship: Scale and Limits," *Journal of Roman Studies* 109: 27–69.

Le Bohec, Yann (1990), *La Sardaigne et l'armée romaine sous le Haut-Empire*, Sassari: Carlo Delfino Editore.

Le Bohec, Yann (2000), *The Imperial Roman Army* (transl. R. Bate), London/New York: Routledge (orig. Paris: Picard, 1989).

Le Bohec, Yann (2006), *L'armée romaine sous le Bas-Empire*, Paris, Picard.

Le Bohec, Yann (2015), "Remarques d'histoire militaire sur la guerre des Maccabées," *Revue des Études Juives*, 174 (1–2): 175–191.

Lee, A. Doug (2007), *War in Late Antiquity. A Social History*, Oxford: Blackwell.

Lefort, Louis (1875), "Cimetière chrétien de Julia Concordia (Porto Gruaro, en Vénétie)," *Revue Archéologique* 29: pp. 340–346.

Lemaire, André (1973), "Le sabbat à l'époque royale israélite," *Revue Biblique* 80: 161–185.

Leon, Harry Joshua (1928), "The Names of the Jews of Ancient Rome," *Transactions and Proceedings of the American Philological Association* 59: 205–224.

Leon, Harry Joshua (1995), *The Jews of Ancient Rome* (update edition with a new introduction by C.A. Osiek), Peabody (Mass.): Hendrickson (orig. 1960).

Lesquier, Jean (1911), *Les institutions militaires de l'Egypte sous les Lagides*, Paris: Leroux.

Letta, Cesare (2020), *Tra umano e divino. Forme e limiti del culto degli imperatori nel mondo romano*, Sarzana/Lugano: Agorà & Co.

Lettich, Giovanni (1983), *Le iscrizioni sepolcrali tardoantiche di Iulia Concordia*, Trieste: Centro Studi Storico-Religiosi Friuli-Venezia Giulia.

Levick, Barbara (1999), *Vespasian*, London/New York: Routledge.

Levine, Lee (2005), *The Ancient Synagogue. The First Thousand Years*, New Haven: Yale University Press (orig. 2000).

Lewis, Naphtali (1991), "The World of P.Yadin," *Bulletin of the American Society of Papyrologists* 28: 35–41.

Lifshitz, Baruch (1977), "Jérusalem sous la domination romaine. Histoire de la ville depuis la conquête de Pompée jusqu'à Constantin (63 a. C.–325 p. C.)," in H. Temporini and W. Haase (eds.), *Aufstieg und Niedergang der römischen Welt, II. Principat*, Berlin/New York: Walter de Gruyter, 444–489.

Lightstone, Jack N. (2011), "Migration and Emergence of Greco-Roman Diaspora Judaism," in Ph. Harland (ed.), *Travel and Religion in Antiquity*, Waterloo (Ontario): Wilfrid Laurier University Press, 187–211.

Llewelyn, Stephen R. (1987), "The Career of T. Mucius Clemens and Its Jewish Connections," in *Idem* (ed.), *New Documents Illustrating Early Christianity, 8. A Review of the Greek Inscriptions and Papyri Published 1984–85*, Grand Rapids: Eerdmans, 1998, 152–155.

Lucas, Leopold (1993), *The Conflict between Christianity and Judaism. A Contribution to the History of the Jews in the Fourth Century* (transl. A.J. Wells), Warminster. Aris & Phillips (orig. Berlin, 1910).

Lukyn Williams, Arthur (1935), *Adversus Judaeos. A Bird's-Eye View of Christian Apologiae until the Renaissance*, Cambridge: Cambridge University Press.

Lüderitz, Gert (1994), "What is the Politeuma?" in J.W. van Henten and P.W. van der Horst (eds.), *Studies in Early Jewish Epigraphy*, Leiden/New York/Köln: Brill, 183–225.

MacAdam, Henry Innes (2002), *Geography, Urbanisation and Settlement Patterns in the Roman Near East*, London/New York: Routledge.

Magness, Jodi (2005), "Heaven on Earth: Helios and the Zodiac Cycle in Ancient Palestinian Synagogues," *Dumbarton Oaks Papers* 59: 1–52.

Mahaffy, John Pentland (1895), *The Empire of the Ptolomies*, London/New York: Macmillan.

Maier, Johann (1992), *Storia del giudaismo nell'antichità* (transl. U. Proch), Brescia: Paideia (orig. Darmstadt: Wissenschaftliche Buchgesellschaft, 1989).

Mann, Michelle L. (2008), "Ethnicity and Identity," in G. Shipley, J. Vanderspoel, D. Mattingly, and L. Foxhall (ed.), *The Cambridge Dictionary of Classical Civilization*, Cambridge: Cambridge University Press, 328–329.

Martin, Alain (1983), "T. Mucius Clemens, *praefectus castrorum* d'Agrippa II," *Zeitschrift für Papyrologie und Epigraphik* 52: 203–210.

Martin, Alain (1985), "À propos de l'inscription de T. Mucius Clemens," *Zeitschrift für Papyrologie und Epigraphik* 60: 275–276.

Maulucci Vivolo, Francesco Paolo (1993), *Pompei. I graffiti figurati*, Foggia: Bastogi Editrice Italiana.

Maxfield, (2009), "'Where Did They Put the Men?' An Enquiry into the Accommodation of Soldiers in Roman Egypt," in W.S. Hanson (ed.), *The Army and Frontiers of Rome: Papers Offered to David J. Breeze on the Occasion of His Sixty-Fifth Birthday and His Retirement from Historic Scotland*, Portsmouth (RI): Journal of Roman Archaeology (Supplementary Series, 74), 63–82.

Mazar, Benjamin (1957), *Beth She'arim. Report on the Excavations during 1936–1940, Volume 1: The Catacombs 1–4*, Jerusalem: The Israel Exploration Society (reed. New Brunswick, NJ: Rutgers University Press, 1973).

Mazzilli, Francesca (2018), *Rural Cult Centres in the Hauran As Part of the Broader Network of the Near East (100 BC–AD 300)*, Oxford: Archaeopress Publishing (Archaeopress Roman Archaeology, 51).

Mazzoleni, Danilo (1975), "Le catacombe ebraiche di Roma," *Studi Romani* 23 (3): 289–302.

Mazzoleni, Danilo (1976), "Les sépultures souterraines des Juifs d' Italie," *Les Dossiers de l'Archéologie* 19: 82–98.

Mazzoleni, Danilo (2002), "L'epigrafia cristiana a Concordia," in *Idem*, *Epigrafi del mondo cristiano antico*, Roma: Lateran University Press, 173–186 (orig. *Antichità Altoadriatiche* 55, 1987: 75–91).

McCasland, S. Vernon (1943), "'Soldiers on Service': The Draft among the Hebrews," *Journal of Biblical Literature* 62: 59–71.

McKay, Heather (1994), *Sabbath and Synagogue. The Question of Sabbath Worship in Ancient Judaism*, Leiden/New York/Köln: E.J. Brill.

Mélèze Modrezejewski, Joseph (1995), *The Jews of Egypt. From Rameses II to Emperor Hadrian* (transl. R. Cornman), Edinburgh: T&T Clark (orig. Paris: Éditions Armand Colin 1992).

Mendelssohn, Ludwig (1875), "*Senatus consulta Romanorum quae sunt in Josephi Antiquitatibus*," *Acta Societatis Philologiae Lipsiensis* 5: 87–288.

Merrill, Elmer Truesdell (1919), "The Expulsion of the Jews from Rome under Tiberius," *Classical Philology* 14: 365–372.

Michael, J. Hugh (1924), "The Jewish Sabbath in the Latin Classical Writers," *American Journal of Semitic Languages and Literatures* 40 (2): 117–124.

Millar, Fergus (1977), *The Emperor in the Roman World*, London: Duckworth.

Millar, Fergus (1993), *The Roman Near East, 31 BC–AD 337*, Cambridge (Mass.): Harvard University Press.

Mimouni, Simon Claude (2007), *La circoncision dans le monde judéen aux époques grecque et romaine. Histoire d'un conflit interne au judaïsme*, Paris/Louvain: Peeters.

Mócsy, András (1974), *Pannonia and Upper Moesia. A History of the Middle Danube Provinces of the Roman Empire* (transl. Sh. Frere), London: Routledge and Kegan Paul (orig. 1916).

Momigliano, Arnaldo (1967), *Ricerche sull'organizzazione della Giudea sotto il dominio Romano, 63 a.C.–70 d.C.*, Amsterdam: Hakkert.

Mor, Menahem (2016), *The Second Jewish Revolt. The Bar Kokhba War, 132–136 CE*, Leiden/Boston: Brill.

Müller, Nicoló (1886), "Le catacombe degli Ebrei presso la via Appia Pignatelli," *Mitteilungen des Deutschen Archaeologischen Instituts. Römische Abteilung* 1: 49–56.

Müller, Nicoló (1915), "Il cimitero degli antichi ebrei posto sulla Via Portuense," *Dissertazioni della Pontificia Accademia Romana di Archeologia* (Ser. II) 12, 202–318.

Murray, Michele (2004), *Playing a Jewish Game. Gentile Christian Judaizing in the First and Second Centuries CE.*, Waterloo (Ontario): Wilfrid Laurier University Press.

Mussies, Gerard, "Jewish Personal Names in Some Non-Literary Sources," in J.W. van Henten and P.W. van der Horst (eds.), *Studies in Early Jewish Epigraphy*, Leiden/New York/Köln: Brill, 242–276.

Myers, Elaine Anne (2010), *The Ituraeans and the Roman Near East. Reassessing the Sources*, Cambridge: Cambridge University Press.

Naville, Edouard (1890), *The Mound of the Jews and the City of Onias. Belbeis, Samanood, Abusir, Tukh el Karmus*, London: Egypt Exploration Fund.

Negev, Avraham (1991), *Personal Names in the Nabatean Realm*, Jerusalem: Hebrew University of Jerusalem.

Neira Falero, Concepción (2005), *La* Notitia Dignitatum. *Nueva edición crítica y comentario histórico*, Madrid: CSIC.

Nelis-Clément, Jocelyne (2006), "Les *stationes* comme espace et transmission du pouvoir," in A. Kolb (ed.), *Herrschaftsstrukturen und Herrschaftspraxis. Konzepte, Prinzipien und Strategien der Administration im römischen Kaiserreich*, Berlin: Akademie, 269–298.

Nemo-Pekelman, Capucine (2008), "L'exclusion des Juifs des fonctions et des dignités," in J.N. Guinot and F. Richard (eds.), *Empire Chrétien et Église aux IVe et Ve siècles. Intégration ou "concordat"? Le témoignage du Code Théodosien*, Paris: Les Éditions du Cerf, 413–430.

Nemo-Pekelman, Capucine (2010), *Rome et ses citoyens juifs (IVe–Ve siècles)*, Paris, Honoré Champion.

Nemo-Pekelman, Capucine (2015), "The Ambiguous Notions of Legal 'statutes' and 'status' in Blumenkranz's work," in Ph. Buc, M. Keil and J. Tolan (eds.), *Jews and Christians in Medieval Europe. The Historiographical Legacy of Bernhard Blumenkranz*, Turnhout: Brepols, 23–34.

Neusner, Jacob (2016), "The Religious Meaning of the Halakhah," in C. Cordoni and G. Langer (eds.), *"Let the Wise Listen and Add to Their Learning" (Prov 1:5). Festschrift for Günter Stemberger on the Occasion of His 75th Birthday*, Berlin/Boston: Walter de Gruyter, 269–290.

Nicasie, Martijn J. (1998), *Twilight to Empire. The Roman Army from the Reign of Diocletian until the Battle of Adrianople*, Amsterdam: J.C. Gieben Publisher, 1998.

Nicolet, Claude (1980), *The World of the Citizen in Republican Rome* (transl. P.S. Falla), Berkeley: Los Angeles University of California Press.

Nikiprowetzky, Valentin (2000), "Le sabbat et les armes dans l'histoire ancienne d'Israël," *Revue des Études Juives* 159 (1–2): 1–17 [posthumous article].

Nock, Arthur Darby (1952), "The Roman Army and the Roman Religious Year," *Harvard Theological Review* 45 (4): 187–252.

Nodet, Étienne (1997), *A Search for the Origins of Judaism. From Joshua to the Mishnah* (transl. E. Crowley), Sheffield: Sheffield Academic Press (orig. Paris 1992).

Nodet, Étienne (2019), "La Galilée au temps de Jésus," in D. Jaffé (dir.), *Juifs et chrétiens aux premiers siècles. Identités, dialogues et dissidences*, Paris: Les Éditions du Cerf, 57–81.

Noethlichs, Karl Leo (1996), *Das Judentum und der römische Staat. Minderheitenpolitik im antiken Rom*, Darmstadt: Wissenschaftliche Buchgesellschaft.

Noethlichs, Karl Leo (2001), *Die Juden im christlichen Imperium Romanum (4.–6 Jahrhundert)*, Berlin: Akademie Verlag.

Noy, David (1994), "The Jewish Communities of Leontopolis and Venosa," in J.W. van Henten and P.W. van der Horst (eds.), *Studies in Early Jewish Epigraphy*, Leiden/New York/Köln: Brill, 162–182.

Olshanetsky, Haggai (2015), "Warriors of Zion. Jewish Soldiers in Hellenistic Armies," *Ancient Warfare* IX (5): 20–24.

Olshanetsky, Haggai (2017), "Israelites in the Army of Assyria in the Name of Gods and Kings," *Ancient Warfare* XI (4): 44–47.

Olshanetsky, Haggai (2018), "Jewish Soldiers of Titus and His Heirs, the Emperors of Rome. Jewish Military Service in the Armies of Rome," *Hayo Haya* (Jerusalem), 13: 7–28 [in Hebrew].

Oppenheimer, Aharon (2005a), "Jewish Conscripts in the Roman Army?" in A. Oppenheimer (ed.), *Between Rome and Babylon. Studies in Jewish Leadership and Society*, Tübingen: Mohr Siebeck, 183–191.

Oppenheimer, Aharon (2005b), "Jewish Penal Authority in Roman Judaea," in A. Oppenheimer (ed.), *Between Rome and Babylon. Studies in Jewish Leadership and Society*, Tübingen: Mohr Siebeck, 173–182.

Oppenheimer, Aharon (2005c), "Sabbatheiligung im Bar-Kochba-Aufstand," in A. Oppenheimer (ed.), *Between Rome and Babylon. Studies in Jewish Leadership and Society*, Tübingen: Mohr Siebeck, 292–302.

Otzen, Benedikt (1990), *Judaism in Antiquity. Political Development and Religious Currents from Alexander to Hadrian* (transl. by F.H. Cryer), Sheffield: Sheffield Academic Press (orig. København: G.E.C. Gad, 1984).

Panayotov, Alexander (2004), "The Jews in the Balkan Provinces of the Roman Empire: The Evidence from the Territory of Bulgaria," in J.M. Barclay (ed.), *Negotiating Diaspora. Jewish Strategies in the Roman Empire*, London/New York: T&T Clark International, 38–65.

Park, Joseph S. (2000), *Conceptions of Afterlife in Jewish Inscriptions. With Special Reference to Pauline Literature*, Tübingen: Mohr Siebeck.

Parker, Robert (1983), *Miasma. Pollution and Purification in Early Greek Religion*, Oxford: Clarendon Press.

Patai, Raphael (1996), *The Jews of Hungary. History, Culture, Psychology*, Detroit (Michigan): Wayne State University Press.

Pavan, Massimiliano (1987), "Concordia tra IV e V secolo," *Antichità Altoadriatiche* 31 (2): 7–28.

Pearce, Sarah (ed.) (2013), *The Image and Its Prohibition in Jewish Antiquity*, Oxford: Journal of Jewish Studies.

Pearce, Sarah (2017), "The Cleopatras and the Jews," *Transactions of the Royal Historical Society* 27: pp. 29–64.

Peppard, Michael (2016), *The World's Oldest Church: Bible, Art, and Ritual at Dura- Europos, Syria*, New Haven/London: Yale University.

Perani, Mauro (2017), "La documentazione sulla presenza ebraica in Sardegna," in A. Foa, G. Lacerenza, and D. Jalla (eds.), *Ebrei, una storia italiana. I primi mille anni*, Milano: Electa, 117–125.

Perea Yébenes, Sabino (2009), "Los diplomas militares: documentos singulares para la integración jurídica y social de los soldados *peregrini* al servicio de Roma. Una introducción a su estudio," in G. Bravo and R. González Salinero (eds.), *Formas de integración en el mundo romano*, Madrid / Salamanca: Signifer Libros, 97–118.

Perea Yébenes, Sabino (2020), *El ejército romano en Egipto*, Madrid: Dilema.

Perea Yébenes, Sabino and Raúl González Salinero (2021), "Urna de *Iulia Sabbathis* (notas a *CIL* VI 20662)," *Epigraphica* 83 (1–2): 349–360.

Pfeiffer, Stefan (2008), "Die alexandrinischen Juden im Spannungsfeld von griechischer Bürgerschaft und römischer Zentralherrschaft. Der Krieg des Jahres 66 n. Chr. in Alexandria," *Klio. Beiträge zur alten Geschichte* 90: 387–402.

Phang, Sara Elise (2012), *Roman Military Service. Ideologies of Discipline in the Late Republic and Early Principate*, Cambridge: Cambridge University Press.

Pflaum, Hans-Georg (1982), *Les carrières procuratoriennes équestres sous le Haut-Empire Romain: Supplement*, Paris: Paul Geuthner.

Pollard, Nigel (1996), "The Roman Army as 'Total Institution' in the Near East? Dura-Europos as a Case Study," in D. Kennedy (ed.), *The Roman Army in the East*, Ann Arbor: Cushing-Malloy Inc., 211–227.

Pollard, Nigel (2000), *Soldiers, Cities and Civilians in Roman Syria*, Michigan: University of Michigan Press.

Porten, Bezalel (1968), *Archives from Elephantine. The Life of an Ancient Jewish Military Colony*, Berkeley/Los Angeles: University of California Press.

Porten, Bezalel (2011), *The Elephantine Papyri in English: Three Millennia of Cross-cultural Continuity and Change* (ed. by Joel Farber *et alii*), Atlanta: Society of Biblical Literature (orig. Leiden: Brill, 1996).

Prag, Jonathan R.W. (2011), "Provincial Governors and Auxiliar Soldiers," in N. Barrandon and Fr. Kirbihler (dirs.), *Les gouverneurs et les provinciaux sous la République romaine*, Rennes: Presses Universitaires de Rennes, 15–28.

Premerstein, Anton von (1903), "Die Buchführung einer agyptischen Legionsabteilung," *Klio. Beiträge zur alten Geschichte* 3: 1–46.

Price, Jonathan J. (1991), "The Enigma of Philip Ben Jakimos," *Historia. Zeitschrift für Alte Geschichte* 40 (1): 77–94.

Price, Jonathan J. (1992), *Jerusalem under Siege. The Collapse of the Jewish State, 66–70 C.E.*, Leiden: Brill.

Prontera, Francesco (ed.) (2009), Tabula Peutingeriana. *Le antiche vie del mondo*, Firenze: Leo S. Olschki Editore (orig. 2003).

Pucci ben Zeev, Miriam (1994), "Marcus Antonius, Publius Dolabella and the Jews," *Athenaeum* 82: 31–40.

Pucci ben Zeev, Miriam (1995), "Caesar and Jewish Law," *Revue Biblique* 102 (1): 28–37.

Pucci ben Zeev, Miriam (1996), "Jewish Rights in the Roman World: New Perspectives," in B. Isaac and A. Oppenheimer (eds.), *Studies on the Jewish Diaspora in the Hellenistic and Roman Periods*, Tel Aviv: Tel-Aviv University/Ramot Publishing, 39–53.

Pucci ben Zeev, Miriam (2002), "Five Delegations to Marcus Antonius (44–41 BCE) and Josephus' Apologetic Purposes", *Materia Giudaica* 7 (1): 2002: 24–27.

Pucci ben Zeev, Miriam (2005), *Diaspora Judaism in Turmoil, 116/117 CE: Ancient Sources and Modern Insights*, Leuven: Peeters.

Pucci ben Zeev, Miriam (2016), "Roman Attitudes to Jews and Judaism in the First Century B.C.E.: Between Hellenistic Traditions and Local Realities," in C. Cordoni and G. Langer (eds.), *"Let the Wise Listen and Add to Their Learning"* (*Prov 1:5*). *Festschrift for Günter Stemberger on the Occasion of His 75th Birthday*, Berlin/Boston: Walter de Gruyter, 2016, 111–125.

Quine, Cat (2019), "The Host of Heaven and the Divine Army: A Reassessment," *Journal of Biblical Literature* 138 (4): 741–755.

Rabello, Alfredo Mordechai (1980), "The Legal Condition of the Jews in the Roman Empire," in H. Temporini (ed.), *Aufstieg und Niedergang der römischen Welt, II, 13. Principat*, Berlin/New York: Walter de Gruyter, 662–762.

Rabello, Alfredo Mordechai (1984), "On the Relations between Diocletian and the Jews," *Journal of Jewish Studies* 35: 147–167.

Rabello, Alfredo Mordechai (1987–1988), *Giustiniano, Ebrei e Samaritani alla luce delle fonti storico-letterarie, ecclesiastiche e giuridiche*, Milano: Giuffrè (2 vols.).

Rabinowich, Nosson Dovid (1988), *The Iggeres of Rav Sherira Gaon*, Jerusalem/New York: Rabbi Jacob Joseph School Press/Ahavath Torah Institute.

Radin, Max (1915), *The Jews among the Greeks and Romans*, Philadelphia: The Jewish Publication Society of America.

Rajak, Tessa (1984), "Was There a Roman Charter of the Jews?," *Journal of Roman Studies* 84: 107–123.

Rajak, Tessa (1985), "Jewish Rights in the Greek Cities under Roman Rule: A New Approach," in W.S. Green (ed.), *Approaches to Ancient Judaism, v. Studies in Judaism and Its Greco-Roman Context*, Atlanta (Georgia): Scholars Press, 19–35.

Rajak, Tessa (1992), "The Jewish Community and Its Boundaries," in J. Lieu, J. North and T. Rajak (eds.), *The Jews among Pagans and Christians in the Roman Empire*, London/New York: Routledge, 9–28.

Rajak, Tessa (1994), "Inscription and Context: Reading the Jewish Catacombs of Rome," in J.W. van Henten and P.W. van der Horst (eds.), *Studies in Early Jewish Epigraphy*, Leiden/New York/Köln: Brill, 226–241.

Rajak, Tessa (1998a), "The Gifts of God at Sardis," in M. Goodman (ed.), *Jews in a Graeco-Roman World*, Oxford: Clarendon Press, 229–239.

Rajak, Tessa (1998b), "The Rabbinic Dead and the Diaspora Dead at Beth She'arim," in P. Schäfer (ed.), *The Talmud Yerushalmi and Graeco-Roman Culture*, Tübingen: Mohr Siebeck, 349–366.

Rajak, Tessa (2005), "Josephus in the Diaspora," in J. Edmondson, St. Mason, and J. Rives (eds.), *Flavius Josephus and Flavian Rome*, Oxford: Oxford University Press, 79–97.

Rajak, Tessa (2013), "The Synagogue Paintings of Dura-Europos: Triumphalism and Competition," in S. Pearce (ed.), *The Image and Its Prohibition in Jewish Antiquity*, Oxford: Journal of Jewish Studies, 89–109.

Rajak, Tessa (2017), "The Mediterranean Jewish Diaspora in the Second Century," in J. Carleton Paget and J. Lieu (eds.), *Christianity in the Second Century. Themes and Developments*, Cambridge: Cambridge University Press, 39–56.

Reichardt, Klaus Dieter (1978), "Die Judengesetzgebung im Codex Theodosianus," *Kairos. Zeitschrift für Religionswissenschaft und Theologie* 20: 16–39.

Reinach, Théodore (1900), "Judaei," in Ch.V. Daremberg and Edm. Saglio (dirs.), *Dictionnaire des antiquités grecques et romaines d'après les textes et les monuments*, Paris: Librairie Hachette et Cie, III.2, 619–632.

Reinach, Théodore (1920), "Review to Jean Juster, *Les Juifs dans l'Empire Romain, leur condition juridique, économique et sociale*. Paris, Geuthner, 1914. In-8°, 2 vol., XII + 510 et VII + 338 pages," *Revue des Études Juives* 70–71: 93–96.

Reynolds, Joyce and Robert Tannenbaum (1987), *Jews and God-Fearers at Aphrodisias. Greek Inscriptions with Commentary*, Cambridge: Cambridge University Press/Cambridge Philological Society.

Rhyder, Julia (2019), "Sabbath and Sanctuary Cult in the Holiness Legislation: A Reassessment," *Journal of Biblical Literature* 138 (4): 721–740.

Ricci, Cecilia (2006), *Stranieri illustri e comunità immigrate a Roma. Vox diversa populorum*, Roma: Edizioni Quasar (Vita e Costumi nel mondo romano antico 28).

Richardson, Peter (1998), "Augustean-Era Synagogues in Rome," in K.P. Donfried and

P. Richardson (eds.), *Judaism and Christianity in First-Century Rome*, Grand Rapids (Michigan)/Cambridge: William B. Eerdmans, 17–29.

Richardson, Peter (1999), *Herod. King of the Jews and Friend of the Romans*, Edinburgh: T&T Clark (orig. University of South Carolina, 1996).

Richmond, Ian Archibald (1962), "The Roman Army and Roman Religion," *Bulletin of the John Rylands Library* 45: 185–197.

Ritter, Bradley (2011), "On the 'πολίτευμα in Heracleopolis,'" *Scripta Classica Israelica* 30: 9–37.

Ritter, Bradley (2015), *Judeans in the Greek Cities of the Roman Empire. Rights, Citizenship and Civil Discord*, Leiden/Boston: Brill.

Ritterling, Emil (1902), "Zur Erklärung von Arrians ἔκταξις κατ' Ἀλανῶν," *Wiener Studien* 24, 359–372.

Rocca, Samuel (2008), *Herod's Judaea. A Mediterranean State in the Classical World*, Tübingen: Mohr Siebeck.

Rocca, Samuel (2010), "Josephus, Suetonius, and Tacitus on Military Service of the Jews of Rome: Discrimination or Norm?," *Italia* 20: 7–30.

Rocca, Samuel (2019), "From *Collegium* to *Ecclesia*: The Changing External Framework of the Jewish Communities in Roman Italy," in K. Berthelot and J. Price (eds.), *In the Crucible of Empire. The Impact of Roman Citizenship upon Greeks, Jews and Christians*, Leuven/Paris/Bristol: Peeters, 217–248.

Rocco, Marco (2012), *L'esercito romano tardoantico. Persistenze e cesure dai Severi a Teodosio I*, Padova: Libreria Universitaria.it Edizioni.

Roddaz, Jean-Michel (1984), *Marcus Agrippa*, Rome: École Française de Rome.

Rodrigues, Nuno Simões (2007), Iudaei in Vrbe. *Os judeus em Roma de Pompeio aos Flávios*, Lisboa/Coimbra: Fundação Calouste Gulbenkian/ Fundação para a Ciência e a Tecnologia.

Rodríguez González, Julio (2001), *Historia de las legiones romanas*, Madrid: Signifer Libros (2 vols.).

Rokéah, David (1995), "Tacitus and Ancient Antisemitism," *Revue des Études Juives* 154 (3–4): 281–294.

Rosemblum, Jordan D. (2016), *The Jewish Dietary Laws in the Ancient World*, Cambridge: Cambridge University Press.

Rosenfeld, Ben Zion and Rivka Potchebutzky (2009), "The Civilian-Military Community in the Two Phases of the Synagogue at Dura Europos: A New Approach," *Levant* 41 (2): 195–222.

Rossi, Anna Maria (1975), "Ricerche sulle multe sepolcrali romane," *Rivista Storica dell'Antichità*, 5: 111–159.

Rostovtzeff, Michael I. (1918), "*Synteleia Tironon*," *Journal of Roman Studies* 8: 26–33.

Rostovtzeff, Michael I. (1943), "*Res Gestae Divi Saporis* and Dura," *Berytus Archaeological Studies* 8: 17–60.

Roth, Cecil (^{2}2007), "Tiberius Julius Alexander", in Fr. Skolnik (ed.), *Encyclopedia Judaica (Second Edition)*, Detroit/Jerusalem: Thomson Gale / Keter, vol. 19, 716–717 (orig. Jerusalem: Kter, 1971, 1135).

Roth, Jonathan P. (1999), *The Logistics of the Roman Army at War (264 B.C.–A.D. 235)*, Leiden/Boston/Köln: Brill.

Roth, Jonathan P. (2007), "Jews and the Roman Army: Perceptions and Realities," in L. de Blois and E. Lo Cascio (eds.), *The Impact of the Roman Army (200 BC–AD 476). Economic, Social, Political, Religious, and Cultural Aspects*, Leiden/Boston: Brill, 409–420.

Roth, Jonathan P. (2019), "Jewish Military Forces in the Roman Service," in Chr. B. Zeichmann (ed.), *Essential Essays for the Study of the Military in First-Century Palestine: Soldiers and the New Testament Context*, Eugene (OR): Pickwick Publications, 79–94.

Roth-Gerson, Leah. (2001), *The Jews of Syria in the Light of the Greek Inscriptions*, Jerusalem: The Zalman Shazar Center (Hebrew).

Roux, Michel (2020), "Les Juifs de Phrygie et les service militaire de l'époque hellénistique à l'époque romaine," *Revue des Études Juives* 179 (1–2): 1–34.

Rubin, Ze'ev (2002), "*Res Gestae Divi Saporis*: Greek and Middle Iranian in a Document of Sasanian Anti-Roman Propaganda," in J.N. Adams, M. Janse, and S. Swain (eds.), *Bilingualism in Ancient Society. Language Contact and the Written Text*, Oxford: Oxford University Press, 267–297.

Rüpke, Jörg (1990), Domi Militiae. *Die religiöse Konstruktion des Krieges in Rom*, Stuttgart: Franz Steiner.

Rutgers, Leonard Victor (1998a), *The Hidden Heritage of Diaspora Judaism*, Leuven: Peeters.

Rutgers, Leonard Victor (1998b), "Roman Policy towards the Jews: Expulsion from the City of Rome during the First Century C.E.", in K.P. Donfried and P. Richardson (eds.), *Judaism and Christianity in First-Century Rome*, Grand Rapids (Michigan)/Cambridge: William B. Eerdmans, 93–116.

Rutgers, Leonard Victor (2000), *The Jews in Late Ancient Rome. Evidence of Cultural Interaction in the Roman Diaspora*, Leiden/Boston/Köln: Brill (orig. 1995).

Rutgers, Leonard Victor (2013), "Cemeteries and Catacombs," in P. Erdkamp (ed.), *Cambridge Companion to Ancient Rome*, Cambridge: Cambridge University Press, 497–521.

Rutgers, Leonard Victor (2016), "'Next Year in Sardis': Reflections on whether the Jewish Diaspora of the Roman Era Was Diasporic at all," in C. Cordoni and G. Langer (eds.), *"Let the Wise Listen and Add to Their Learning" (Prov 1:5). Festschrift for Günter Stemberger on the Occasion of His 75th Birthday*, Berlin/Boston: Walter de Gruyter, 167–195.

Sablayrolles, Robert (1996), Libertinus miles. *Les cohortes de vigiles*, Rome: École Française de Rome.

Saddington, Dennis Bain (1982), *The Development of the Roman Auxiliary Forces from Caesar to Vespasian (49 B.C.–A.D. 79)*, Harare: University of Zimbabwe.

Saddington, Dennis Bain (1995), "Problems in Military Ranks and Military Personnel in Josephus," in Y. Le Bohec (ed.), *La hiérarchie (Rangordnung) de l'armée romaine sous le Haut-Empire*, Paris: De Boccard, 53–55.

Saddington, Dennis Bain (2009), "Client Kings' Armies under Augustus: The Case of Herod," in D.M. Jacobson and N. Kokkinos (eds.), *Herod and Augustus*, Leiden/Boston: Brill, 303–323.

Safrai, Samuel (1971), "The Relations between the Roman Army and the Jews of Eretz Yisrael after the Destruction of the Second Temple," in *Roman Frontier Studies 1967. The Proceedings of the Seventh International Congress Held at Tel Aviv*, Tel Aviv: Tel Aviv University, 224–230.

Sänger, Patrick (2019), *Die ptolemäische Organisationsform* politeuma. *Ein Herrschaftsinstrument zugunsten jüdischer und anderer hellenischer Gemeinschaften*, Tübingen: Mohr Siebeck (Texts and Studies in Ancient Judaism 178).

Sartre, Maurice (1991), *L'orient romain: provinces et sociétés provinciales en Méditerranée orientale d'Auguste aux Sévères (31 avant J.-C.–235 après J.-C.)*, Paris: Seuil.

Sartre, Maurice (2005), *The Middle East under Rome* (transl. C. Porter and E. Rawlings), Cambridge (Mass.): Harvard University Press.

Sartre, Maurice (2011), "Brigands, colons et pouvoirs en Syrie du Sud au Ier siècle de notre ère," *Anabases. Traditions et Réceptions de l'Antiquité* 13: 1–35 (printed version, 207–245).

Saulnier, Christiane (1981), "Lois romaines sur les Juifs selon Flavius Josèphe," *Revue Biblique* 88: 161–198.

Savvopoulos, Kyriakos (2018), "New Data concerning the Foundation of Alexandria," in Chr. S. Zerefos and M.V. Verdinoyannis (eds.), *Hellenistic Alexandria. Celebrating 24 Centuries*, Oxford: Archaeopress Publishing, 115–132.

Schäfer, Peter (1997), *Judeophobia. Attitudes toward the Jews in the Ancient World*, Cambridge/London: Harvard University Press.

Schalit, Abraham (1960), "The Letter of Antiochus III to Zeuxis regarding the Establishment of Jewish Military Colonies in Phrygia and Lydia," *Jewish Quarterly Review* 50: 289–318.

Scharf, Ralf (1997), "*Regii Emeseni Iudaei*: Bemerkungen zu einer spätantiken Truppe," *Latomus* 56: 343–359.

Schimanowski, Gottfried (2007), "Die jüdische Integration in die Oberschicht Alexandriens und die angebliche Apostasie des Tiberius Julius Alexander," in J. Frey, D.R. Schwartz, and St. Gripentrog (eds.), *Jewish Identity in the Greco-Roman World. Jüdische Identität in der griechisch-römischen Welt*, Leiden/Boston: Brill, 111–135.

Schoenfeld, Andrew J. (2006), "Sons of Israel in Caesar's Service: Jewish Soldiers in the Roman Military," *Shofar. An Interdisciplinary Journal of Jewish Studies* 24 (3): 115–126.

Schürer, Emil (1973), *The History of the Jewish People in the Age of Jesus Christ (175 BC–AD 135), vol. I* (transl. and ed. by G. Vermes, F. Millar, M. Black and P. Vermes), Edinburgh: T&T Clark (orig. 1886).

Schürer, Emil (1979), *The History of the Jewish People in the Age of Jesus Christ (175 BC–AD 135), vol. II* (transl. and ed. by G. Vermes, F. Millar, M. Black and P. Vermes), Edinburgh: T&T Clark (orig. 1887).

Schürer, Emil (1986), *The History of the Jewish People in the Age of Jesus Christ (175 BC–AD 135), vol. III, 1* (transl. and ed. by G. Vermes, F. Millar, M. Black, P. Vermes and M. Goodman), Edinburgh: T&T Clark (orig. 1890).

Schürer, Emil (1987), *The History of the Jewish People in the Age of Jesus Christ (175 BC–AD 135), vol. III, 2* (transl. and ed. by G. Vermes, F. Millar, M. Black, P. Vermes and M. Goodman), Edinburgh: T&T Clark (orig. 1890).

Schwartz, Daniel R. (2009), "Flavio Giuseppe," in F. Coarelli (ed.), *Divus Vespasianus. II bimillenario dei Flavi*, Milano: Electa, 34–41.

Schwartz, Daniel R. (2013), "Judeans, Jews and Their Neighbors: Jewish Identity in the Second Temple Period," in R. Albert and J. Wöhrle (eds.), *Between Cooperation and Hostility. Multiple Identities in Ancient Judaism and the Interaction with Foreign Powers*, Göttingen: Vandenhoeck & Ruprecht, 13–31.

Schwartz, Seth (1984), "T. Mucius Clemens, Commander of the Army of Agrippa II: An Epigraphical Note," *Zeitschrift für Papyrologie und Epigraphik* 56: 240–242.

Schwartz, Seth (1985) "T. Mucius Clemens: A Supplementary Note," *Zeitschrift für Papyrologie und Epigraphik* 58: 296.

Schwartz, Seth (1990), *Josephus and Judaean Politics*, Leiden: Brill.

Schwartz, Seth (2001), *Imperialism and Jewish Society (200 B.C.E to 640 C.E.)*, Princeton: Princeton University Press.

Schwartz, Seth (2002), "Rabbinization in the Sixth Century," in P. Schäfer (ed.), *The Talmud Yerushalmi and Graeco-Roman Culture, III*, Tübingen: Mohr Siebeck, 55–69.

Schwartz, Seth (2016), "Jewish Communities in the Roman Diaspora: Why Salo Baron Still Matters?" in Y. Furstenberg (ed.), *Jewish and Christian Communal Identities in the Roman World*, Leiden/Boston: Brill, 225–242.

Schwartz, Seth (2017), "Law in Jewish Society in the Second Temple Period," in Chr. Hayes (ed.), *The Cambridge Companion to Judaism and Law*, Cambridge: Cambridge University Press, 48–75.

Seeman, Chris (2013), *Rome and Judea in Transition: Hasmonean Relations with the Roman Republic and the Evolution of the High Priesthood*, New York: Peter Lang.

Sekunda, Nick (1994), *Seleucid and Ptolemaic Reformed Armies 168–145 BC, Vol. 1. The Seleucid Army under Antiochus IV Epiphanes*, Stockport (Cheshire): Montvert.

Sevenster, Jan Nicolaas (1975), *The Roots of Pagan Anti-Semitism in the Ancient World*, Leiden: Brill.

Seward, Desmond (2009), *Jerusalem's Traitor. Josephus, Masada, and the Fall of Judea*, Cambridge, MA: Da Capo Press.

Seyrig, Henri (1965), "Antiquités syriennes," *Syria* 42 (1/2): 25–34.

Shatzman, Israel (1991), *The Armies of the Hasmonaeans and Herod. From Hellenistic to Roman Frameworks*, Tübingen: J.C.B. Mohr (Paul Siebeck).

Shatzman, Israel (1999), "The Integration of Iudaea into the Roman Empire," *Scripta Classica Israelica* 18: 49–84.

Shatzman, Israel (2012), "The Expansionist Policy of John Hyrcanus and his Relations with Rome," in G. Urso (ed.), *Iudaea socia, Iudaea capta. Atti del Convegno Internazionale Cividale del Friuli (22–24 settembre 2011)*, Pisa: Edizioni ETS, 29–77.

Shean, John F. (2010), *Soldering for God. Christianity and the Roman Army*, Leiden/Boston: Brill.

Sherk, Robert K. (1969), *Roman Documents from the Greek East. Senatus Consulta and Epistulae to the Age of Augustus*, Baltimore/Maryland: The Johns Hopkins Press.

Sherwin-White, Adrian Nicholas (1984), *Roman Foreign Policy in the East, 168 B.C. to A.D. 1*, London: Duckworth.

Sherwin-White, Adrian Nicholas (1999), *The Roman Citizenship*, Oxford: Oxford University Press (orig. 1939).

Smallwood, Edith Mary (1956), "Some Notes on the Jews under Tiberius," *Latomus* 15: 314–329.

Smallwood, Edith Mary (1976), *The Jews under Roman Rule. From Pompey to Diocletian*, Leiden: Brill.

Smallwood, Edith Mary (1999), "The Diaspora in the Roman Period Before CE 70," in W. Horbury, W.D. Davies, and J. Sturdy (eds.), *The Cambridge History of Judaism, III. The Early Roman Period*, Cambridge: Cambridge University Press, 168–191.

Smith, Richard Edwin (1958), *Service in the Post-Marian Roman Army*, Manchester: Manchester University Press.

Solin, H (1983), "Juden und Syrer im westlichen Teil der römischen Welt. Eine ethnisch-demographische Studie mit besonderer Berücksichtigung der sprachlichen Zustände," in W. Haase (ed.), *Aufstieg und Niedergang der römischen Welt, II. Principat, 29.2*, Walter de Gruyter, Berlin/New York, 587–789.

Southern, Pat (1989), "The *Numeri* of the Roman Imperial Army," *Britannia* 20: 81–140.

Southern, Pat (2007), *The Roman Army: A Social and Institutional History*, Oxford: Oxford University Press.

Southern, Pat and Karen R. Dixon (2000), *The Late Roman Army*, London/New York: Routledge (orig. 1996).

Spaul, John E.H. (1994), *Ala²: The Auxiliary Cavalry Units of the Pre-Diocletianic Imperial Roman Army*, Andover: Nectoreca Press.

Spaul, John E.H. (2000), *Cohors²: The Evidence for and a Short History of the Auxiliary*

Units of the Imperial Roman Army, Oxford: Archaeopress Publishing (British Archaeological Reports, 841).

Speidel, Michael Alexander (1992), "Roman Army Pay Scales," *Journal of Roman Studies* 82, 87–106.

Speidel, Michael Alexander and Hans Lieb (eds.) (2007), *Militärdiplome: die Forschungsbeiträge der Berner Gespräche von 2004*, Stuttgart: Franz Steiner.

Speidel, Michael P. (1975), "The Rise of Ethnic Units in the Roman Imperial Army," in H. Temporini (ed.), *Aufstieg und Niedergang der römischen Welt, II.3. Politische Geschichte. Provinzen und Randvölker (Allgemeines; Britannien, Hispanien, Gallien)*, Berlin/New York: Walter de Gruyter, 202–231.

Speidel, Michael P. (1977), "The Roman Army in Arabia," in H. Temporini and W. Haase (eds.), *Aufstieg und Niedergang der römischen Welt, II.8. Principat*, Berlin/New York: Walter de Gruyter, 687–730 (= *Idem* 1984, *Roman Army Studies*, Amsterdam: J.C. Gieben 1984, I, 229–272).

Speidel, Michael P. (1982–1983), "The Roman Army in Judaea under the Procurators. The Italian and the Augustan Cohort in the Acts of the Apostles," *Ancient Society* 13–14: 233–240.

Speidel, Michael P. (1996), "Raising New Units for the Late Roman Army: 'Auxilia Palatina'," *Dumbarton Oaks Papers* 50: 163–195.

Spera, Lucrezia (1999), *Il Paesaggio suburbano di Roma dall'Antichità al Medioevo. Il comprensorio tra le vie Latine ed Ardeatina dalle mura Aureliane al III miglio*, Roma: "L'Erma" di Bretschneider, 1999.

Stemberger, Günter (2000), *Jews and Christians in the Holy Land. Palestine in the Fourth Century* (transl. R. Tuschling), Edinburgh: T&T Clark (orig. München, 1987).

Stemberger, Günter (2009), *Das klassische Judentum. Kultur und Geschichte der rabbinischen Zeit*, München: C.H. Beck.

Stern, Karen B. (2008), *Inscribing Devotion and Death. Archaeological Evidence for Jewish Populations of North Africa*, Leiden/Boston: Brill.

Stern, Karen B. (2013), "Graffiti as Gift: Mortuary and Devotional Graffiti in the Late Ancient Levant," in M. Satlow (ed.), *The Gift in Antiquity*, New York: Wiley/Blackwell, 137–157.

Stern, Karen B. (2018), *Writing on the Wall. Graffiti and the Forgotten Jews of Antiquity*, Princeton/Oxford: Princeton University Press.

Stern, Sacha (1995), "The Concept of Authorship in the Babylonian Talmud," *Journal of Jewish Studies*, 46, 183–195.

Syme, Ronald (1970), "Domitius Corbulo," *Journal of Roman Studies* 60: 27–39.

Talbert, Richard J.A. (ed.) (2000), *Barrington Atlas of the Greek and Roman World*, Princeton: Princeton University Press.

Talmon, Shemaryahu (1991), "The Internal Diversification of Judaism in the Early Second Period," in Sh. Talmon (ed.), *Jewish Civilization in the Hellenistic-Roman Period*, Sheffield: Sheffield Academic Press, 16–43.

Tcherikover, Victor (1982), *Hellenistic Civilization and the Jews*, New York: Atheneum (orig. Forge Village, Mass. 1959).

Terry, Rhoda and Steven Fine, "Mosaic Depiction of a Roman Soldier," in St. Fine (ed.), *Sacred Realm. The Emergence of the Synagogue in the Ancient World*, New York/Oxford: Oxford University Press/Yeshiva University Museum, 166–167.

Tibor, Grüll (2016), "Jewish Presence in the Danubian Provinces of the Roman Empire," in M. Hrbácsek (ed.), *Židovský kurltúrny fenomén v stredoeurópskom kontexte. Zsidó kultúra közép-európai kontextusban*, Nitra: UKF, 9–18.

Tomlin, Roger S.O. (1972), "*Seniores-iuniores* in the Late-Roman Field Army," *American Journal of Philology* 93: 253–278.

Tomlin, Roger S.O. (1976), "*Notitia Dignitatum omnium, tam civilium quam militarium*," in R. Goodburn and P. Bartholomew (eds.), *Aspects of the Notitia Dignitatum*, Oxford: BAR, 189–209.

Tomlin, Roger S.O. (1998), "Christianity and the Late Roman Army." in S.N.C. Lieu and D. Montserrat (eds.), *Constantine. History, Historiography and Legend*, London/New York: Routledge, 21–51.

Tonstad, Sigve K. (2016), "To Fight or not to Fight: The Sabbath and the Maccabean Revolt," *Andrew University Seminary Studies* 54 (1): 135–146.

Trebilco, Paul R. (1991), *Jewish Communities in Asia Minor*, Cambridge: Cambridge University Press.

Troiani, Lucio (2012), "Le operazioni militari di Pompeo in Giudea dalla visuale di Giuseppe," in G. Urso (ed.), *Iudaea socia, Iudaea capta. Atti del Convegno Internazionale Cividale del Friuli (22–24 settembre 2011)*, Pisa: Edizioni ETS, 89–95.

Tropper, Amran (2016), *Rewriting Ancient Jewish History. The History of the Jews in Roman Times and the New Historical Method*, London/New York: Routledge.

Tuval, Michael (2013), *From Jerusalem Priest to Roman Jew. On Josephus and the Paradigms of Ancient Judaism*, Tübingen: Mohr Siebeck.

Tzaferis, Vassilios (1982), "A Monumental Roman Tomb on Tell 'Eitun," *'Atiqot* 8: 22–25 [Hebrew].

Ureche, Petru (2017), "*Numeri.* Contingentes 'especiales' en el ejército romano," *Desperta Ferro* 13: 56–61.

Valbelle, Dominique and Jean-Yves Carrez-Maratray (2001), *Le camp romain du Bas-Empire à Tell El-Herr*, Paris: Editions Errance.

van der Horst, Pieter Willen (1991), *Ancient Jewish Epitaphs. An Introductory Survey of a Millennium of Jewish Funerary Epigraphy (300 BCE–700 CE)*, Kampen: Kok Pharos.

van der Horst, Pieter Willen (2013), "Judaism in Asia Minor," in M.R. Salzman and W. Adler (eds.), *The Cambridge History of Religions in the Ancient World, II. From the Hellenistic Age to Late Antiquity*, Cambridge: Cambridge University Press, 321–340 (= *Idem* (2014), *Studies in Ancient Judaism and Early Christianity*, Leiden/Boston: Brill).

van der Horst, Pieter Willen (2015), Saxa judaica loquuntur: *Lessons from Early Jewish Inscriptions. Radboud Prestige Lectures 2014*, Leiden/Boston: Brill.

van der Lans, Birgit (2015), "The Politics of Exclusion. Expulsion of Jews and Others from Rome," in M. Labahn and O. Lehtipuu (eds.), *People under Power. Early Jewish and Christian Responses to the Roman Empire*, Amsterdam: Amsterdam University Press, 33–78.

Varon, Perlina (2001), "Testimonianze del servizio prestato dagli ebrei nell'esercito romano," in A. Lewin (ed.), *Gli ebrei nell'Impero romano. Saggi vari*, Firenze: Giuntina, 271–277.

Vismara, Cinzia (1986), "I cimiteri ebraici di Roma," in A. Giardina (ed.), *Società romana e Impero tardoantico, II. Roma, politica, economia, paesaggio urbano*, Roma: Laterza, 352–392 and 490–503.

Vitale, Micaela (1986), "Le catacombe ebraiche: premesse per un recupero," *La Rassegna Mensile di Israel* 52 (1): 176–186.

Vogelstein, Hermann and Paul Rieger, *Geschichte der Juden in Rom*, Berlin: Mayer & Müller, 1895–1896 (2 vols.).

Vogler, Chantal (1979), "Les Juifs dans le Code Théodosien," in J. Le Brun (ed.), *Les chrétiens devant le fait juif. Jalons historiques*, Paris: Beauchesne, 35–74.

Watson, George Ronald (1985), *The Roman Soldier. Aspects of Greek and Roman Life*, Ithaca (New York): Cornell University Press (orig. London: Thames and Hudson, 1969).

Webster, Graham (1985), *The Roman Imperial Army*, Totowa (N.J.): Barnes & Noble.

Weisman, Stefanie (2012), "Militarism in the Wall Paintings of the Dura-Europos Synagogue: A New Perspective on Jewish Life on the Roman Frontier," *Shofar. An Interdisciplinary Journal of Jewish Studies* 30 (3): 1–34.

Weiss, Herold (1991), "Philo on the Sabbath," in D.T. Runia, D.M. Hay and D. Winston (eds.), *Heirs of the Septuagint: Philo, Hellenistic Judaism and Early Christianity. Festschrift for Earle Hilgert*, Atlanta: Society of Biblical Literature, 83–105.

Weiss, Ze'ev (2010), "Burial Practices in Beth She'arim and the Question of Dating the Patriarchal Necropolis," in Z. Weiss, O. Irshai, J. Magness, and S. Schwartz (eds.), *"Follow the Wise". Studies in Jewish History and Culture in Honor of Lee I. Levine*, Winona Lake (Indiana): Eisenbrauns, 207–231.

Weiss, Ze'ev (2013), "Images and Figural Representations in the urban Galilee: Defining Limits in Times of Shifting Borders," in S. Pearce (ed.), *The Image and Its Prohibition in Jewish Antiquity*, Oxford: Journal of Jewish Studies, 130–144.

Weiss, Ze'ev (2014), *Public Spectacles in Roman and Late Antique Palestine*, Cambridge (MA)/London: Harvard University Press.

Weitzman, Steven (2017), *The Origin of the Jews. The Quest for Roots in a Rootless Age*, Princeton: Princeton University Press.

Welles, Bradford *et alii* (1959), *The Parchments and Papyri: Final Report 5, Part 1 of The Excavations at Dura-Europos* (ed. A. Perkins), New Haven: Yale University Press.

Wendt, Heidi (2015), "*Iudaica Romana*. A Rereading of Judean Expulsions from Rome," *Journal of Ancient Judaism* 6: 97–126.

Whelan, Robin (2018), "Mirrors for Bureaucrats: Expectations of Christian Officials in the Theodosian Empire," *Journal of Roman Studies* 108: 74–98.

Wilde, Robert (1949), *The Treatment of the Jews in the Greek Christian Writers of the First Three Centuries*, Washington D.C.: The Catholic University of America Press.

Wilker, Julia (2007), "*Herodes Iudaicus*. Herodes als 'jüdischer König'," in L.-M. Günther (ed.), *Herodes und Rom*, Stuttgart: Franz Steiner, 2007, 27–45.

Wilker, Julia (2008), "Unabhängigkeit durch Integration. Zu den jüdisch-römischen Beziehungen im 2. Jahrhundert v.Chr.," in M. Karrer and W. Kraus (eds.), *Die Septuaginta—Texte, Kontexte, Lebenswelten*, Tübingen: Mohr Siebeck, 194–201.

Wilker, Julia (2018), "Between Empires and Peers: Hasmonean Foreign Policy under Alexander Jannaeus," *Electrum* 25: 127–145.

Williams, Margaret H. (1989), "The Expulsion of the Jews from Rome in A.D. 19," *Latomus* 48: 764–784.

Williams, Margaret H. (1997), "The Meaning and Function of *Ioudaios* in Graeco-Roman Inscriptions," *Zeitschrift für Papyrologie und Epigraphik* 116: 249–262.

Williams, Margaret H. (1998a), "The Structure of the Jewish Community in Rome," in M. Goodman (ed.), *Jews in a Graeco-Roman World*, Oxford: Clarendon Press, 215–228.

Williams, Margaret H. (1998b), *The Jews among the Greeks and Romans: A Diasporan Sourcebook*, London: Duckworth.

Williams, Margaret H. (2013), *Jews in a Graeco-Roman Environment*, Tübingen: Mohr Siebeck.

Wise, Michael Owen (2015), *Language and Literacy in Roman Judaea. A Study of the Bar Kokhba Documents*, Yale University Press, New Haven/London.

Woods, David (1992), "A Note concerning *Regii Emeseni Iudaei*," *Latomus* 51: 404–407.

Worp, Klaas A. (1986), "Four Greek Ostraka from the Thermenmuseum (Heerlen)," *Zeitschrift für Papyrologie und Epigraphik* 65: 191–194.

Wright, Benjamin J., III (2015), *The Letter of Aristeas. 'Aristeas to Philocrates' or 'On the Translation of the Law of the Jews'*, Berlin/Boston: Walter de Gruyter.

Yavetz, Zvi (1997), *Judenfeindschaft in der Antike. Die Münchener Vorträge* (Introd. Chr. Meier), München: C.H. Beck.

Zack, Andreas (2018), "C. Iulius Caesar und Hyrkanus II. Überlegungen zur chronologischen Abfolge der Dokumente bei Flavius Josephus *ant.* 14,10,2–10 (190–222) und *ant.* 18,8,5 (145–148)," *Electrum* 25: 147–185.

Zeichmann, Christopher B. (2015), "Herodian Kings and Their Soldiers in the Acts of the Apostles: A Response to Craig Keener," *Journal of Greco-Roman Christianity and Judaism* 11: 178–190.

Zeichmann, Christopher B. (2018), *The Roman Army and the New Testament*, Lanham/Boulder/New York/London: Lexington Books/Fortress Academic.

Zerbini, Andrea (2016), "Human Mobility in the Roman near East: Patterns and Motives," in L. de Ligt and L.E. Tacoma (eds.), *Migration and Mobility in Early Roman Empire*, Leiden/Boston: Brill, 305–344.

Zollschan, Linda (2017), *Rome and Judaea. International Law Relations, 162–100 BCE*, London/New York: Routledge.

Zuccotti, Ferdinando (2000), *Il giuramento nel mondo giuridico e religioso antico. Elementi per uno studio comparatistico*, Milano: Giuffrè.

Zuckerman, Constantin (1993), "Les "barbares" romans: au suject de l'opinion des *auxilia* tétrarchiques," in F. Vallet and M. Kazanski (eds.), *L'armée romain et les Barbares du IIIe au VIIe siécle*, Paris: Association Française d'Archéologie Mérovingienne / Musée des Antiquités Nationales, 17–20.

Analytical Index

www.ingramcontent.com/pod-product-compliance
Ingram Content Group UK Ltd.
Pitfield, Milton Keynes, MK11 3LW, UK
UKHW021421091026
15624UKWH00010B/332

9 789004 506756